SpringerBriefs in Applied Sciences and Technology

Computational Intelligence

Series Editor

Janusz Kacprzyk, Systems Research Institute, Polish Academy of Sciences, Warsaw, Poland

SpringerBriefs in Computational Intelligence are a series of slim high-quality publications encompassing the entire spectrum of Computational Intelligence. Featuring compact volumes of 50 to 125 pages (approximately 20,000-45,000 words), Briefs are shorter than a conventional book but longer than a journal article. Thus Briefs serve as timely, concise tools for students, researchers, and professionals.

Soumi Majumder · Bitan Misra

Analysing Trends and Patterns in Employee Engagement Through AI

Soumi Majumder
Department of Business Administration
Future Institute of Engineering
and Management
Kolkata, West Bengal, India

Bitan Misra
Department of Computer Science
and Engineering
Techno International New Town
Kolkata, West Bengal, India

ISSN 2191-530X ISSN 2191-5318 (electronic)
SpringerBriefs in Applied Sciences and Technology
ISSN 2625-3704 ISSN 2625-3712 (electronic)
SpringerBriefs in Computational Intelligence
ISBN 978-981-96-4495-7 ISBN 978-981-96-4496-4 (eBook)
https://doi.org/10.1007/978-981-96-4496-4

© The Editor(s) (if applicable) and The Author(s), under exclusive license to Springer Nature Singapore Pte Ltd. 2025

This work is subject to copyright. All rights are solely and exclusively licensed by the Publisher, whether the whole or part of the material is concerned, specifically the rights of translation, reprinting, reuse of illustrations, recitation, broadcasting, reproduction on microfilms or in any other physical way, and transmission or information storage and retrieval, electronic adaptation, computer software, or by similar or dissimilar methodology now known or hereafter developed.
The use of general descriptive names, registered names, trademarks, service marks, etc. in this publication does not imply, even in the absence of a specific statement, that such names are exempt from the relevant protective laws and regulations and therefore free for general use.
The publisher, the authors and the editors are safe to assume that the advice and information in this book are believed to be true and accurate at the date of publication. Neither the publisher nor the authors or the editors give a warranty, expressed or implied, with respect to the material contained herein or for any errors or omissions that may have been made. The publisher remains neutral with regard to jurisdictional claims in published maps and institutional affiliations.

This Springer imprint is published by the registered company Springer Nature Singapore Pte Ltd.
The registered company address is: 152 Beach Road, #21-01/04 Gateway East, Singapore 189721, Singapore

If disposing of this product, please recycle the paper.

Preface

When it comes to evaluating employee engagement data and seeing important trends and patterns, AI can be quite helpful. AI has the power to improve corporate results and change the employee experience. AI keeps workers engaged and focused by providing personalised learning and real-time accomplishment recognition. Artificial Intelligence (AI) is revolutionising employee engagement and talent retention through personalised learning experiences and optimised workflows. In general, human resources managers can leverage AI as a potent tool to better understand their employees through information and develop more strategic, fact-based decisions that will increase employee engagement. Today, it is possible to fully utilise AI's potential through employee involvement; it is no more an unrealistic dream.

The book covers different aspects of AI application in employee engagement. Chapter 1 provides a brief introduction to various roles and responsibilities of HR, senior leaders, managers and employees. The importance of employee engagement trends in organisations is presented in Chap. 2. Chapter 3 discusses various challenges related to employee engagement. The significance of employee engagement levels in workplace are highlighted in Chap. 4. Various employee engagement models are discussed in Chap. 5. Application of Artificial Intelligence (AI) in employee engagement and retention is presented in Chap. 6. Chapter 7 discusses the concept of leveraging AI in employee engagement. Chapter 8 discusses various AI techniques and methodologies to analyse employee data for various purpose related to employee engagement and retention. Chapter 9 presents a predictive analysis on employing AI in engagement. Chapter 10 discusses about the advantages, weaknesses and future scope of AI-empowered employee engagement. The last section draws multiple conclusions about the book.

Kolkata, India

Soumi Majumder
Bitan Misra

Competing Interests The authors have no competing interests to declare that are relevant to the content of this manuscript.

Contents

1	**Introduction**		1
	1.1	The Role and Responsibilities of Senior Leaders in Employee Engagement	2
		1.1.1 Exemplify Required Behaviour	3
		1.1.2 Define Goals and Plans	3
		1.1.3 Encourage Initiatives	3
		1.1.4 Thoughtful Communication	3
	1.2	The Role and Responsibilities of HR in Employee Engagement	4
		1.2.1 Encourage Responsibility and Alignment	4
		1.2.2 Selection of Effective Processes and Tools	4
		1.2.3 Mentor Leaders and Teams	4
	1.3	The Role and Responsibilities of Managers in Employee Engagement	5
		1.3.1 Establish a Secure Work Environment	5
		1.3.2 Evaluate, Discuss and Act Upon the Outcome	5
		1.3.3 Take Responsibility for Action	5
		1.3.4 Engage in Behaviour that Drives Engagement	6
	1.4	The Role and Responsibilities of Employees in Employee Engagement	6
		1.4.1 Suggestions Should Be Freely Shared	6
		1.4.2 Generate a List of Potential Solutions	6
		1.4.3 Encourage Adherence to Promises	7
	References		7

2	**Importance of Measuring Employee Engagement Trends in Organisations**		9
	2.1	Employee Well-Being	10
	2.2	Employee Development	10
	2.3	Effective Communication	11
	2.4	Leadership Development	11
	2.5	Team Collaboration	11
	2.6	Diversity and Inclusion	12
	2.7	Employee Advocacy	12
	References		12
3	**Employee Engagement Challenges**		13
	3.1	Excessive Red Tape	13
	3.2	Line Managers and Supervisors with Poor Communication Skills	14
	3.3	Lack of One-on-One Time	14
	3.4	Shortage of Growth Opportunities	14
	3.5	Not Enough Transparency	15
	3.6	Employees Who Feel Unheard	15
	3.7	Lack of Alignment	15
	3.8	Lack of Recognition	16
	3.9	Lack of Direction	16
	3.10	Lack of Balance	16
	References		17
4	**Significance of Employee Engagement Levels in the Workplace**		19
	4.1	Highly Engaged Employees	19
	4.2	Moderately Engaged Employees	20
	4.3	Barely Engaged Employees	20
	4.4	Disengaged Employees	21
	4.5	Employee Retention	21
	4.6	Employee Productivity	22
	4.7	Increased Profitability	22
	4.8	Less Burnout and Better Mental Wellness	22
	4.9	Reduction in Absenteeism	22
	References		23
5	**Employee Engagement Models**		25
	5.1	The Zinger Model	25
	5.2	The Gallup Model	26
	5.3	The AON-Hewitt Model	27
	5.4	The Kahn Model	27
	5.5	The Maslow Model	28
	5.6	The JD-R Model	28

6	**Artificial Intelligence in Employee Engagement and Retention**		31
	6.1	Career Planning	34
	6.2	Identifying Opportunities for Growth	34
	6.3	Better Work-Life Balance	34
	6.4	Achieve Equitable Compensation	34
	6.5	Improving Productivity	35
	References		35
7	**Leveraging AI for Employee Engagement**		37
	7.1	AI for Employee Recognition and Rewards	38
		7.1.1 Key Steps to Develop Effective Automated Recognition Systems	38
		7.1.2 Advantages of Automated Systems	39
		7.1.3 The Challenges of Automated Systems Include	39
	7.2	Gamification and AI-Powered Incentives	39
		7.2.1 Personalisation	40
		7.2.2 Predictive Insights	40
		7.2.3 Testing and Improvement	40
		7.2.4 Dynamic Content	41
		7.2.5 Fraud Prevention	41
		7.2.6 Comments Analysis	41
		7.2.7 AI-Based Player Segmentation Targets Specific Gamification Tactics and Rewards to Increase Player Engagement	41
	7.3	AI-Enabled Performance Management	41
		7.3.1 Automated Performance Review	42
		7.3.2 Seamless Data Collection	42
		7.3.3 Reduced Bias	42
		7.3.4 Real-Time Data Analysis	43
	7.4	Enhancing Employee Communication and Collaboration with AI	43
		7.4.1 Translating and Customisation	44
		7.4.2 Improving Cooperation	44
		7.4.3 Personalisation	44
		7.4.4 Streamlining Regular Conversations	44
		7.4.5 Commercial Interactions	44
	7.5	Intelligent Virtual Assistants	45
		7.5.1 Natural Language Processing (NLP)	45
		7.5.2 Responsive Learning and Predictive AI	45
		7.5.3 Generative AI	46
	7.6	Augmenting Decision-Making with AI	46
	7.7	Improving Employee Development and Learning with AI	46
	7.8	Virtual Reality and AI for Immersive Training	46

	7.9	Measure and Manage Employee Sentiment	47
	7.10	Improving Personalisation	47
	7.11	Obtaining Support from Leadership	47
		References	47
8	**AI-Powered Analytics Uncover Hidden Patterns in Employee Engagement Data**		51
	8.1	Natural Language Processing (NLP)	51
		8.1.1 NLP in Sentiment Analysis	52
		8.1.2 NLP in Employee Recruitment	53
		8.1.3 NLP in HR Chatbots	53
		8.1.4 NLP in Employee Engagement	54
	8.2	Machine Learning	54
		8.2.1 Encourage Learning	55
		8.2.2 Promotion of Cooperation	55
		8.2.3 Boost Well-Being	55
		8.2.4 Exploring Novel Concepts and Being Creative	56
		8.2.5 Suggest Feedback	56
	8.3	GenAI	56
		References	58
9	**Predictive Analytics: Anticipating Future Engagement Trends with AI**		61
	9.1	Case Study—Applications of Predictive Analytics in Marketing Management	62
		9.1.1 Customer Segmentation	62
		9.1.2 Churn Prediction	62
		9.1.3 Lead Scoring	63
		9.1.4 Content Personalisation	63
	9.2	Predictive Analytics Tools for Marketers	64
	9.3	Caselets on Employee Engagement in Various Organisations	64
		9.3.1 Google: The Incentives for Employee Engagement	64
		9.3.2 Zappos: Enlightening an Energetic Company Culture	65
		9.3.3 Microsoft: A Data-Driven Attitude	66
		9.3.4 Hilton: Gratitude and Obligation	66
		9.3.5 The Ritz Carlton: Endowing Employees to Drive the Additional Mile	67
		References	68
10	**Ethical Considerations and Future Directions of AI in Employee Engagement**		69
	10.1	Ethical Use of AI for Employee Monitoring	69
	10.2	Ensuring Human-AI Collaboration and Empathy	70
	10.3	Exploring Emerging Trends and Innovations in AI for Employee Engagement	70

10.4	AI Integration with Current Systems	71
10.5	Making AI Ethical and Objective	71
10.6	Constant Monitoring and Assessment	72
10.7	Handling Information Security and Morality	72
10.8	Handling of Employment Issues	72
10.9	Expansion of Execution	72
10.10	Responsibility and Disclosure	73
10.11	Low Emotional Intelligence	73
10.12	Antiquated Facilities	73
10.13	Complexity of the Learning Curve	73
10.14	Organisational Culture Shift	73
10.15	Combining Services for Generative AI	74
References		74

Conclusions 75

About the Authors

Soumi Majumder completed her PGDM (specialization in Human Resource Management) from the All India Management Association, Ministry of HRD, Government of India, 2012. She did her Diploma in Labour Laws with Administrative Laws (DLLAL) and Master of Business Administration (HRM) from Annamalai University under the Tamil Nadu Government in 2013 and 2020, respectively. She is an associate researcher at the Universidad Internacional de La Rioja, Logroño, La Rioja, Spain. She is a research scholar in the Department of Management Studies at Lincoln University, Malaysia. She is an assistant professor at the Future Business School, Future Institute of Engineering and Management, Kolkata, India. Previously, she was associated with Sister Nivedita University, Techno India College of Technology, NSHM College of Management and Technology, J. D. Birla Institute of Science and Commerce, West Bengal State Labor Institute, Siliguri, DAITM, and many more recognized B Schools.

Bitan Misra is currently working as an assistant professor, in Department of CSE, Techno International New Town, Kolkata, India. She received her B.Tech. and M.Tech. dual degree in Electronics and Telecommunication Engineering from KIIT University, Bhubaneswar, India in 2018. She received her Ph.D. in 2022 from National Institute of Technology, Durgapur, India. She received a Gold Medal during her UG for securing the highest CGPA in the university. She has published almost more than 30 research papers in various international journals and conferences and has multiple copyright and patents. Her main research interests include optimization techniques, deep learning, evolutionary algorithms, and soft computing techniques. She has worked as a reviewer in several national and international journals and conferences. She is an associate editor of *International Journal of Ambient Computing and Intelligence*, IGI Global. She is a member of IEEE and Internet Society.

Chapter 1
Introduction

The HR concept of employee engagement describes the level of enthusiasm and commitment that a worker feels for his or her job. The engaged workforce cares about their work and the company's performance, believing that it is making a difference. In view of the obvious links between job satisfaction and employee morale, engagement by employees may be essential for a company's success. Communication is a key factor in creating and maintaining employee engagement. Workers who are engaged are more likely to be productive and more efficient. They are also more committed to the values and objectives of the company. There are many ways in which employers can foster employee engagement, for example, through clear communication of expectations and the granting of incentives and promotions to a job well done, ensuring that employees know about company performance and regularly provide feedback. In addition, it is important to provide employees with appreciation and respect as well as the perception that their ideas are being heard and taken care of. Engaged employees believe that their work is meaningful and that they are appreciated and supported.

In the modern pace of business, one may wonder exactly what the role of artificial intelligence (AI) is. AI is like a wizard who's helping us find the hidden secrets of the workforce. Engaged employees are the lifeblood of any successfully operated company. They are passionate souls who bring passion to their job, fuel innovation and drive productivity up. AI can analyse large volumes of employee data quickly and accurately, with advanced algorithms and machine learning playing a significant role in this process [1]. AI can crunch the numbers and detect patterns that the human eye may miss; it works from employee performance metrics to feedback on employees' attitudes or social interactions. In general, the impact of employee engagement on business performance and profitability tends to be decisive at the company level. However, it also provides leaders with an understanding of the needs of their employees as well as identifying ways to improve morale and make working environments more pleasant. Both formal and informal communication are critical factors for fostering a workplace with high employee engagement. It is important

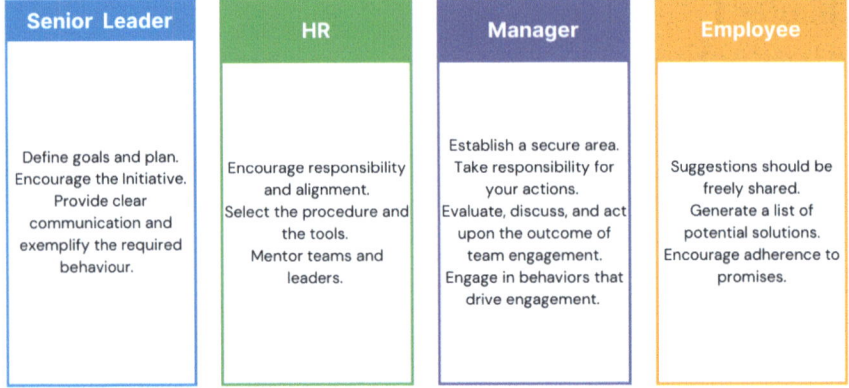

Fig. 1.1 Roles and responsibilities in employee engagement

for a person to obtain tools and services that allow people to connect with each other with as little friction as possible.

However, to improve a company's situation, it is also necessary for the company to receive useful and impartial feedback. HR is frequently required to determine employee engagement in a wide range of organisations. However, the HR cannot and should not have ownership of an employee's engagement. It is important that everyone in the organisation understands and drives engagement, including senior leaders, HR, managers and employees. This concept is illustrated in Fig. 1.1. HR plays a mission-critical role. To make work better, they need to empower everyone else to play their part [2]. The right processes, tools, communication, and training are needed to achieve this. To create a thriving workplace culture, however, everyone in the organisation will need to work together.

1.1 The Role and Responsibilities of Senior Leaders in Employee Engagement

The advocates of employee engagement are the leaders of the organisation. They are the most powerful influences of an organisation's culture. They influence every part of a company's attitude towards engagement, setting the tone for the rest of the company. The rest of the organisation will follow if leaders prioritise engagement. The roles and responsibilities of senior leaders are discussed below.

1.1 The Role and Responsibilities of Senior Leaders in Employee Engagement

1.1.1 *Exemplify Required Behaviour*

An open and enthusiastic attitude towards employee engagement, feedback or change must be demonstrated by senior management. As time goes on, the leadership's attitude towards engagement becomes more and more crucial. An annual survey of engagement should be conducted at a time. Senior leaders should model engagement-focused behaviours such as coaching, regular recognition, feedback and action planning [3].

1.1.2 *Define Goals and Plans*

To develop a long-term strategy for the engagement of employees, leaders should collaborate closely with human resources. They should contribute to the definition of the "why" of engagement and to the link between engagement and business priorities. This will help everyone in the organisation to see how employee engagement efforts have an impact.

1.1.3 *Encourage Initiatives*

Unless senior leaders invest their time, energy, and resources into engagement efforts, no one else will. The identification of strategic focus areas and the sharing of how they expect the organisation to invest in them should be facilitated by senior leaders. While many of the engagement results are to be used to fuel department and team action plans, senior leaders should have the organisation's overall priorities in mind. Through their time, budget, resources and communication, they should support these priorities.

1.1.4 *Thoughtful Communication*

Communication is the starting point for everything that follows in regard to employee engagement. In addition, senior leaders must be at the forefront of communications. By making engagement an important topic of continuous conversation, leaders can help drive accountability and actions across the organisation [4]. By sharing what and why and continuing to keep the organisation informed of progress, leaders should communicate changes and changes in strategy.

1.2 The Role and Responsibilities of HR in Employee Engagement

To align leaders, managers and staff with the aim of engaging and putting strategies into practice, human resources play a key role. The "how" of employee engagement should be taken into account by the HR. They are helping to ensure that everything goes smoothly, and they work together with teams in understanding and driving the measures. The roles and responsibilities of HR in employee engagement are explained below.

1.2.1 Encourage Responsibility and Alignment

To obtain input, alignment and acquire information from all stakeholders within an organisation, an HR should work at every level of the organisation. With senior leaders, this means helping to define the "why" behind engagement and connecting engagement results to key business priorities and KPIs. With managers and employees, it means holding teams accountable for action and the next steps [5].

1.2.2 Selection of Effective Processes and Tools

HR is behind the expert in employee engagement strategies and initiatives that are already being implemented. The goal of HR is to make engagement a successful, year-round, valuable effort for everyone involved [6]. This includes, for example, selecting the easiest and most efficient employee engagement software that will meet teams where they are, implementing an employee engagement survey and other employee feedback processes, building a communication strategy in partnership with senior leaders and managers, creating training programmes and resources related to employee engagement and new technology, supporting teams to analyse their results of engagement and working with teams to implement targeted and effective measures.

1.2.3 Mentor Leaders and Teams

In regard to managing employee engagement, human resources should be a cheerleader for leaders, managers and teams. When engagement issues arise, they should be a source of information that will help other people feel more confident in their capacity to deal with obstacles. In providing them with the necessary tools and resources, HR should encourage managers to engage in this process.

1.3 The Role and Responsibilities of Managers in Employee Engagement

The key role of managers is to act on the ground with their staff and to be closely involved in employee engagement. In addition to serving as a sounding board for employee feedback and ideas, managers should help to promote the initiatives of the organisation as a whole. They should take responsibility for the results of their team engagement and lead them to a better future.

1.3.1 Establish a Secure Work Environment

If employees do not believe that their feedback will be listened to and taken into account, they will refrain from making it publicly available. A safe space for healthy dialogue and open feedback should be created by managers. A large number of managers lack the necessary skills for receiving and responding to feedback, so this should be a developmental priority for HR staff. Open dialogue on the engagement of staff throughout the year is aimed at achieving this.

1.3.2 Evaluate, Discuss and Act Upon the Outcome

The team should be guided by managers who understand and discuss the results of team engagement [7]. These discussions provide opportunities for teams to dig deeper into their results and consider how they can move forward. Managers should assist the team in translating their findings into action plans and monitoring and communicating their progress to the team.

1.3.3 Take Responsibility for Action

The results and progress of the team are ultimately a manager's responsibility. Responsibility for results, actions and progress should be assigned to them. While managers should empower individual contributors to be part of the action planning process, they should not put action back on the team's shoulders. Great managers make action planning for progressing important tasks and not to do more.

1.3.4 Engage in Behaviour that Drives Engagement

Areas of engagement that are not under the manager's responsibility will probably be reflected in team engagement results. Managers should focus on what they can personally be held accountable to while also taking into account the needs of human resources and senior management. However, managers should have a clear understanding of organisational priorities and do what they can support and inform their teams about these efforts.

1.4 The Role and Responsibilities of Employees in Employee Engagement

Without genuine feedback from the front lines, there is no way of knowing if an employee engagement strategy is actually working. The reason why initiatives are introduced and how they will make a difference should be understood by employees [8]. Employees should give considerable thought to how they might share their thoughts and enhance their experience.

1.4.1 Suggestions Should Be Freely Shared

Providing feedback on employees' experiences at work is the most important contribution to employee engagement. The employees were asked to complete an employee questionnaire about concerns, care and honesty. However, they do not have to wait until the annual survey of their engagement to share their views. Through open and continuous dialogue with their managers, teams and leaders, employees should start to share their feedback outside of surveys.

1.4.2 Generate a List of Potential Solutions

To bring tangible ideas and solutions, employees need to take on a broader role than merely sharing their opinions. They should be actively involved in the input on what action to take, with a view to discuss ways of improving staff's experience.

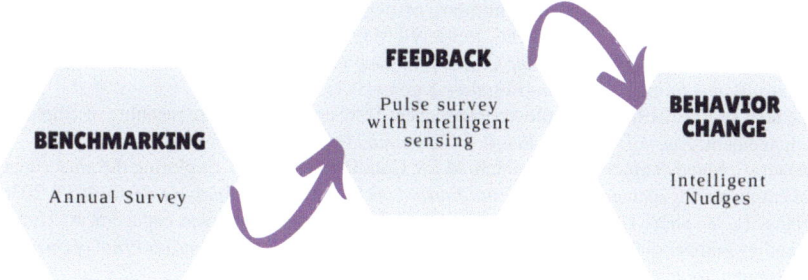

Fig. 1.2 Measuring employee engagement with technology

1.4.3 *Encourage Adherence to Promises*

Understanding employee involvement while also considering the needs of a team is important [9]. The engagement of the team should be clearly understood by the staff, and their role in the overall picture should be taken into account. They should familiarise themselves with what and why of the team's action plan and do their part in supporting the team's broader commitments throughout the year.

Today, organisations are investing time, money and energy in gathering employee engagement data through a series of channels (Fig. 1.2). In view of the most recent advances in technology, which make it much simpler to access employee feedback, AI cannot be ignored for these various forms of data collection on interactions [10]. Organisations must remember that the best way of measuring employee engagement with new technology differs from one organisation to another. In this way, artificial intelligence can be combined with the measurement of employee engagement in a number of effective ways. The emotional side of employee engagement can also be understood by AI [11]. Artificial intelligence can measure general mood, determine possible problems and provide invaluable insight into improvement through the analysis of employee feedback and sentiment in communication with staff. Imagine a crystal ball that tells individuals which employees are at risk of leaving the company or even being fired. Using artificial intelligence, one can act on a proactive basis and offer personalised solutions and support to maintain the motivation and loyalty of the best talent, but not all AI can also help organisations craft targeted employee development programmes, identify trends and skill gaps, and recommend training opportunities tailored to individual needs.

References

1. Nikolova, I., Schaufeli, W., & Notelaers, G. (2019). Engaging leader–engaged employees? A cross-lagged study on employee engagement. *European Management Journal, 37*(6), 772–783.

2. https://www.quantumworkplace.com/future-of-work/employee-engagement-roles-whos-responsible-for-employee-engagement/. Retrieved May 09, 2024.
3. Sun, L., & Bunchapattanasakda, C. (2019). Employee engagement: A literature review. *International Journal of Human Resource Studies, 9*(1), 63–80.
4. Lemon, L. L. (2019). The employee experience: How employees make meaning of employee engagement. *Journal of Public Relations Research, 31*(5–6), 176–199.
5. Alam, J., Mendelson, M., Ibn Boamah, M., & Gauthier, M. (2023). Exploring the antecedents of employee engagement. *International Journal of Organisational Analysis, 31*(6), 2017–2030.
6. Jiang, H., & Shen, H. (2023). Toward a relational theory of employee engagement: Understanding authenticity, transparency, and employee behaviors. *International Journal of Business Communication, 60*(3), 948–975.
7. Turner, P., & Turner, P. (2020). What is employee engagement? In *Employee engagement in contemporary organisations: Maintaining high productivity and sustained competitiveness* (pp. 27–56).
8. Noercahyo, U. S., Maarif, M. S., & Sumertajaya, I. M. (2021). The role of employee engagement on job satisfaction and its effect on organisational performance. *Jurnal Aplikasi Manajemen, 19*(2), 296–309.
9. Moletsane, M., Tefera, O., & Migiro, S. (2019). The relationship between employee engagement and organisational productivity of sugar industry in South Africa: The employees' perspective. *African Journal of Business & Economic Research, 14*(1).
10. https://www.spiceworks.com/hr/hr-strategy/articles/definitive-guide-to-ai-employee-engagement/. Retrieved May 10, 2024.
11. Hughes, C., Robert, L., Frady, K., & Arroyos, A. (2019). Artificial intelligence, employee engagement, fairness, and job outcomes. In *Managing technology and middle-and low-skilled employees: Advances for economic regeneration* (pp. 61–68). Emerald Publishing Limited.

Chapter 2
Importance of Measuring Employee Engagement Trends in Organisations

It is looking at the soul of the organisation when one can measure trends in employee engagement. It provides insight into how employees truly feel about their jobs, colleagues and the whole company culture. This knowledge is invaluable because it enables dealing with pain points, improving team dynamics and creating an atmosphere of optimism in the workplace. The point is that measuring trends in employee engagement is not just about feeling good [1]. It is going to have a direct impact on the bottom line. Employees who are engaged are more likely to go above and beyond, which results in increased productivity and innovation. They are also more likely to remain in business, reducing turnover costs and ensuring the continuity of trade. People can identify areas where the organisation is highly successful and reproduce these successes through the identification of trends in relationships. On the other hand, people can also identify areas where improvements are needed, such as leadership communication, recognition programmes and initiatives to address work-life balance. With this knowledge, one can develop specific strategies to increase employee engagement and create a workplace that is conducive to the well-being of employees. Measuring engagement trends is not a one-and-done deal. This is an ongoing process that will allow monitoring of progress over a period of time. Individuals can assess the effectiveness of their initiatives and make data-based decisions to ensure that employee engagement continues to increase through a comparison of figures from different time periods. Therefore, measuring employee engagement trends is not just a "nice-to-have"—it is a strategic imperative. It enables the construction of a culture that attracts and retains the best talent, and it drives business success to keep the organisation at the forefront of innovativeness. Before diving into analysis, one must specify his/her goals clearly. The analysis will be guided by clear objectives, and one will have access to relevant information.

Choosing the right artificial intelligence tools is also necessary. The selection of AI tools and platforms that are aligned with an organisation's needs and capabilities is highly necessary. To maximise the potential of analyses, one should look for features such as sentiment analysis, natural language processing and predictive analytics. The

© The Author(s), under exclusive license to Springer Nature Singapore Pte Ltd. 2025
S. Majumder and B. Misra, *Analysing Trends and Patterns in Employee Engagement Through AI*, SpringerBriefs in Computational Intelligence,
https://doi.org/10.1007/978-981-96-4496-4_2

data quality is important because the accuracy of AI analysis is directly influenced by the quality of the data. These data must be clear, uniform and complete [2]. To ensure reliable results, removing duplicate data, addressing missing data and validating data are important, i.e., data visualisation techniques. It is important to use the data to highlight outcomes in an easy and persuasive way. Visualisations make it easier for stakeholders to understand complex information and facilitate their decisions. Context is important, while artificial intelligence provides valuable insight. Employee engagement is an ongoing journey. AI can be used for regular monitoring of engagement metrics, as well as to identify new trends. The adjustment of organisational strategies to take into account evolving information and ensure constant improvement and adaptation is highly appreciable. Combining quantitative and qualitative data analysis means that AI is good at analysing quantitative data, but it should not forget the importance of qualitative data. To gain a comprehensive view of employees' experiences, AI analysis using quantitative methods such as focus groups or interviews is considered. It is about using AI as a powerful tool to inform decisions; change the way one do business and develop an engaged and productive workforce [3]. In regard to measuring the benefits of employee engagement, companies should consider the striking benefits of AI-empowered employee engagement. The benefits are discussed below.

2.1 Employee Well-Being

The measurement of employee engagement provides an indication of possible sources of stress, burnout or dissatisfaction among staff. Improving overall well-being, mental health, and work-life balance can contribute to addressing these issues. AI-powered behavioural health services are gaining attraction as a private, easily accessible tool for managing employees' emotional wellness. AI systems that watch and identify human emotions, such as Affectiva, use machine learning. This allows people to keep an eye on their emotions and avoid burnout. The well-known meditation app Headspace uses artificial intelligence (AI) to personalise meditation sessions, which helps users reduce stress and enhance their general well-being.

2.2 Employee Development

The opportunities for growth and development are more likely to be sought by employees who are engaged. Organisations can identify individuals who are motivated to learn and provide them with tailored development plans, enhance their skills or increase their career advancement by quantifying their level of involvement. When intelligent technologies such as machine learning and natural language processing are used in employee training, the learning process is improved. By enabling the development of customised and adaptable training programmes, these technologies increase the difficulty of learning.

2.3 Effective Communication

To assess the effectiveness of internal communication strategies, it is possible for organisations to measure their engagement. The results of engagement can help improve communication channels and ensure that relevant information is delivered to employees on time and in an appropriate manner. The field of artificial intelligence (AI) has completely changed how humans interact. Artificial intelligence has facilitated faster, easier, and more efficient communication through developments in machine learning and natural language processing. Most people define artificial intelligence (AI) as a machine's capacity to mimic or imitate human conduct. Cognitive AI, or generative AI that can mimic human conduct, is most frequently utilised in the communications industry.

2.4 Leadership Development

It is possible to promote leadership development by involving staff in the measurement process. A culture of accountability, transparency and effective management practices is promoted through the encouragement of leaders to be active participants in a strategy for employee engagement and initiative. Organisations may find, evaluate and nurture the most qualified applicants for leadership roles with the use of AI-powered leadership development programmes [4]. AI can also be used to evaluate information from performance reviews and employee surveys to learn more about how workers feel about their existing leaders. By applying these insights, companies may enhance their programmes for developing leaders and ensure that they are providing their staff members with the assistance they need.

2.5 Team Collaboration

Insight into the dynamics of team collaboration is provided by the measurement of engagement. The identification of areas where further support or intervention is needed to strengthen teamwork and cooperation can be achieved by knowing the level of engagement in teams. In regard to working together in the workplace, AI's strongest suit is its capacity for data interpretation. AI has the ability to track team members' actions, allowing for proactive team dynamics improvement. It can introduce a completely new collaborative paradigm into contemporary businesses. By bridging geographical divides, AI-powered virtual collaboration solutions also facilitate productive collaboration among remote teams. Artificial intelligence (AI) capabilities for tasks such as transcription, translation, sentiment analysis of recorded calls, and voice recognition for automated actions can also be integrated into business communication tools, such as virtual PBX (private branch exchange) systems.

2.6 Diversity and Inclusion

The measurement of engagement allows organisations to assess the level of inclusivity at work. Organisations can take targeted action to improve the diversity, fairness and inclusion of their environment by identifying possible differences in participation between various demographic groups.

2.7 Employee Advocacy

Brand ambassadors are more likely to be highly engaged workers who speak positively about their organisation and promote it to others. The measurement of employee engagement helps determine which employees are passionate advocates, while organisations can leverage their influence to improve corporate reputations and attract new talent [5]. AI-based content creation and personalised incentive programmes are beneficial. Generative AI has the potential to be a financially advantageous partner for social employee advocacy initiatives.

References

1. Rao, S., Chitranshi, J., & Punjabi, N. (2020). Role of artificial intelligence in employee engagement and retention. *Journal of Applied Management-Jidnyasa,* 42–60.
2. Braganza, A., Chen, W., Canhoto, A., & Sap, S. (2021). Productive employment and decent work: The impact of AI adoption on psychological contracts, job engagement and employee trust. *Journal of Business Research, 131,* 485–494.
3. Prentice, C., & Nguyen, M. (2020). Engaging and retaining customers with AI and employee service. *Journal of Retailing and Consumer Services, 56,* 102186.
4. Dutta, D., Mishra, S. K., & Tyagi, D. (2023). Augmented employee voice and employee engagement using artificial intelligence-enabled chatbots: A field study. *The International Journal of Human Resource Management, 34*(12), 2451–2480.
5. Braganza, A., Chen, W., Canhoto, A., & Sap, S. (2022). Gigification, job engagement and satisfaction: The moderating role of AI enabled system automation in operations management. *Production Planning & Control, 33*(16), 1534–1547.

Chapter 3
Employee Engagement Challenges

Beyond its trendy facade, employee engagement is the cornerstone of a thriving workplace, weaving a tapestry of productivity and job satisfaction. Understanding these two types lays the foundation for addressing the intricate challenges that may arise on the engagement front. A research study has shown that the turnover rate of a company could be increased by 25–59% with highly motivated employees. Identifying obstacles to high employee engagement is at the heart of all this. These are as follows.

3.1 Excessive Red Tape

To perform basic tasks in the workplace, many procedures and rules should comply with an employee. Many procedures have an adverse impact on staff engagement. Unnecessary rules and procedures are the most common obstacles to employee productivity. According to the Chartered Institute of Personnel and Development (CIPD), the work environment should provide pleasant working experience so that workers can perform their jobs easily and under minimal pressure. As a result, work processes will need to be simplified as much as possible. The Chartered Institute of Personnel and Development (CIPD) is a proficient figure for HR and people enlargement.

3.2 Line Managers and Supervisors with Poor Communication Skills

It is the middle manager who is in a position to bring the skills and focus of the team into line with the organisation's objectives. For example, line managers form an important link between front-line workers and senior managers in factories and other similar establishments. However, they often lack the skills to interact with their teams [1]. The management empowerment report reveals that only one-third of front-line managers received specific training to support employees. Most of them have been selected for this position because they are good workers, not because they are great managers. To address the challenges of employee engagement, inspiration and transparency in communication are essential improvements that can be made by leaders.

3.3 Lack of One-on-One Time

Only 21% of millennials and 18% of non-millennials meet with managers once a week. They say they're only meeting a few times per month. One-on-one is essential for managers to be able to meet with employees on a personal level and to discuss their needs, concerns, growth and possibilities [2]. The key driver of employee engagement could be these factors. The ability of managers to communicate effectively with their staff on projects and on performance has been severely limited without frequent one-on-one meetings. One needs to ensure that the managers of the company meet regularly with their staff to avoid obstacles to good working relationships.

3.4 Shortage of Growth Opportunities

According to the Blessing White study, one of the main reasons why workers leave their jobs is a lack of growth opportunities. Increasing opportunities for development is necessary if anyone wants to remove barriers to employee engagement. The feeling of being stuck in a dead-end job will arise if workers are not aware of their career path, and sooner or later, they are going to jump ship and find better opportunities. On the other hand, employers who encourage their employees to learn new skills not only retain them longer but also benefit from the new knowledge they bring to the table.

3.5 Not Enough Transparency

According to one study, 75% of employees are interested in their employer's performance, and only 23% believe that they have an understanding of how the company operates. A genuine, two-way openness of communication between workers and management is defined as transparency in the workplace. In addition, a lack of transparency is a barrier to employee engagement, which can undermine trust between employees and managers. It is therefore important to show the staff the bigger picture of how their work ties in with corporate objectives as a whole. This results in a high level of communication and trust, thereby contributing to employee productivity and engagement.

3.6 Employees Who Feel Unheard

A survey showed that 82% of employees had ideas for improving their businesses. However, it is difficult for more than one-third of staff to obtain such inputs at the highest level. This is one of the key challenges in terms of employee engagement since employees need to be listened to if they want to take part in their jobs. According to the 2020 employee experience study, organisations that act on feedback have twice the engagement rate of those that do not. Only when senior managers listen to workers' thoughts and opinions can this kind of engagement occur [3].

3.7 Lack of Alignment

The inevitable isolation that their work entails is one of the main challenges in building and maintaining employee engagement with remote workers. It is too easy for business leaders to focus on what is right in front of them in regard to creating a sense of belonging; employees and workers see every day. It is a genuine concern to miss the working culture. Employees must be able to understand how and why their efforts have an impact on the organisation at large. One of the key drivers of employee engagement is alignment with company objectives. Without them, remote workers are at particular risk of feeling isolated rather than being part of a greater effort to achieve success.

3.8 Lack of Recognition

One of the most important issues that needs to be considered when dealing with a remote workforce is feeling unappreciated. The good compatibility of the reward system with all employees must be ensured. A company's in-house team will love an early Friday night finish and drink, but remote workers may feel left out and isolated. This may have a large impact on staff engagement and retention. All reward schemes provide every member of the team with clear and measurable benefits, regardless of their location. In addition to being a transparent reward system, it is also important to make regular recognition an everyday part of the working culture. With in-house staff, this can be achieved easily with verbal praise, perhaps a comment in the hallway, a daily briefing for the staff, or acknowledgement of a job well done [4]. However, there is a need for more effort to be made in the case of remotely employed workers. In this context, written recognition may be useful, such as a weekly email summarising staff achievement or a formal programme such as the "Employee of the Month". This means that those off-site are still a part of the team, and their achievements and accomplishments are actively acknowledged and recognised.

3.9 Lack of Direction

There is a good body of evidence on the impact of leadership on employee engagement, which is where a person truly needs to lead from the front line. It is becoming clear that for remote workers, strong and effective leadership is even more essential. According to the Perceptyx report, only 42% of employees strongly believe that leaders are effectively leading their organisations through this crisis. This means that most people do not think they have a good leader at the helm. A significant barrier to the engagement of staff will be such a lack of confidence.

3.10 Lack of Balance

The precursor to low levels of engagement is a lack of balance between work and personal life. It has always been a challenge to ensure that employees manage their working lives in healthy ways, but the problem is even more acute for remote workers. A person will go a long way towards increasing motivation and productivity if he can help his domestic staff deal with these conceptual issues of employee engagement. A healthy balance between work and life is not conducive to remote work. According to Buffer, 22% of remote workers say they are having trouble disconnecting after work. In addition, a study by Forbes showed that 38% of workers suffer from exhaustion during the entire week of virtual meetings, and 30% are under stress.

References

1. Krishnan, L. R. K. (2023). Employee engagement driven by AI and ML and mediated by progressive work practices—An IT industry perspective.
2. Ahiwale, A., & Bhand, N. S. (2022). Role of artificial intelligence in employee engagement—An exploratory analysis. *IBMRD's Journal of Management & Research,* 158–164.
3. Dhir, S., & Shukla, A. (2019). Role of organisational image in employee engagement and performance. *Benchmarking: An International Journal, 26*(3), 971–989.
4. Shinde, G. R., Majumder, S., Bhapkar, H. R., & Mahalle, P. N. (2022). Exploratory data analysis. In *Quality of work-life during pandemic: Data analysis and mathematical modelling* (pp. 97–105).

Chapter 4
Significance of Employee Engagement Levels in the Workplace

The most successful companies recognise the importance of their staff, knowing that they are a great asset to them. Employee engagement opens the potential of employees, increases their productivity and fuels sustained business growth when they are encouraged and involved. There has been a growing emphasis on employee engagement over the last few years. While most executives believe that engaged employees are performing more effectively, less than half of the respondents said they had seen a favourable return on investment from their initiatives to engage staff, and 37% thought it was an important priority for their organisation. It is vital for organisations to understand the key principles of employee engagement to overcome that gap. It is vital to understand what drives involvement and how it can be effectively measured. This knowledge is essential for creating a working environment in which the involvement of employees not only exists but also significantly contributes to their success. Figure 4.1 shows the levels of employee engagement in the workplace [1]. The pyramid is divided into three levels of engagement: job engagement, which includes (i) passion, (ii) encouragement. Group engagement level consists of (i) dedication, (ii) voluntary work activity. The institutional engagement refers to (i) Desire to continue. The various levels of employee engagement and their significance are discussed below.

4.1 Highly Engaged Employees

The views of highly engaged employees are very favourable to their workplaces. These "brand advocates" speak highly of their company to family and friends. They are encouraging other employees around them to do their best. In a survey of 984 business leaders conducted by Quantum Workplace with the Harvard Business Review, 81% strongly agreed that highly engaged workers are more effective and productive than their average or low engagement.

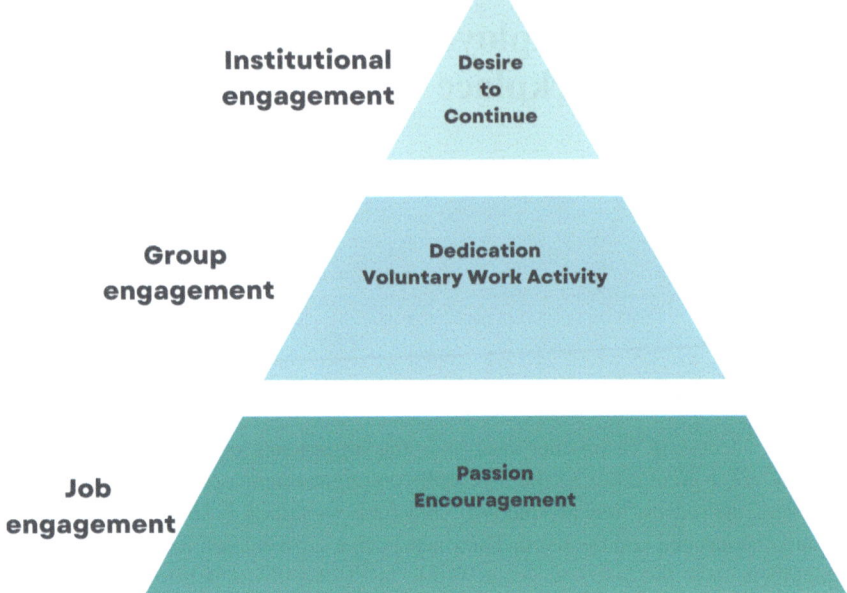

Fig. 4.1 Levels of employee engagement in the workplace

4.2 Moderately Engaged Employees

The organisation is perceived positively by moderately engaged employees. They are fond of their company, but there's something about their organisation, their team or their job that is holding them back from full engagement. They are less likely to demand more responsibility and may not be able to perform as well.

4.3 Barely Engaged Employees

Workers who are seldom engaged feel indifferent about their place of employment. They are typically not motivated to do their job and are doing what they can, sometimes at a lesser rate. Barely engaged employees may be researching other jobs and are at high risk of turnover.

4.4 Disengaged Employees

Disengaged staff members have negative opinions about their jobs. They are not connected to the organisation's mission, objectives or future. They are not committed to their roles and responsibilities. Understanding how to address disengaged workers so that their negative perceptions do not affect the productivity of the workforce around them is important [2].

For engaged employees who are happy and completely committed to their work, it is more than just a paycheck—it is the dedication to their employers and role that makes them passionate about their work, which is often reflected in business success and employee engagement.

To build strong working relationships between workers and improve productivity, employee communication remains a key instrument. Companies can quickly build trust with their workers by offering transparent and accurate communication. Often, because executives believe that an engaged workforce has a direct connection to salary and remuneration, companies miss key issues relating to employee engagement. The majority of leaders believe that the lack of promotion opportunities and a desire to find new jobs are why workers leave their companies. The three most important factors for the growth of a number of organisations, both inside and outside the organisation, are usually connected, communication and engagement. This view contradicts the findings of some recent studies showing that other important factors have a direct effect on employee participation. The following include employee trust in the company's leadership, employee relationships with management teams and supervisors, and employee pride in being a part of the organisation.

4.5 Employee Retention

HR leaders are focused on increasing employee engagement initiatives within their organisations, with one of the key reasons being retention. High employee engagement reduces both turnover and hiring costs, and employee disengagement is a major factor contributing to high employee turnover. To retain top talent, employee engagement is critical and an important element in the puzzle of staff satisfaction, as disengaged workers are more likely to leave their jobs. According to Forbes, employees who are engaged in their work are more likely to be motivated and to remain loyal to their employer. This will make it possible to achieve more business objectives and help drive the organisation forward.

4.6 Employee Productivity

Employees who are engaged at work are more likely to be consistently productive, which leads to a higher-performing workforce. According to TechJury, companies with a higher level of employee engagement are 21% more profitable. The workplace research foundation revealed that employees who are engaged have a 38% greater probability of being more productive than average.

4.7 Increased Profitability

According to Gallup, highly engaged teams have a 21% increase in profitability. It is enough for leaders to see the return on investment in employee engagement strategies and implement them. Compared to business units without engaged workers, businesses with engaged employees have a 23% increase in profit. In addition, teams with workers who thrive can see significant reductions in absenteeism, turnover and accidents as well as greater loyalty to their customers. Engaged employees make it a point to show up to work and do more work—highly engaged business units realise a 41% reduction in absenteeism and a 17% increase in productivity.

4.8 Less Burnout and Better Mental Wellness

It is finally time for mental health and exhaustion to take centre stage in the business world, at a good time. The COVID-19 pandemic has only exacerbated the stress levels of employees, with companies now seeing a record 70% burnout rate and nearly half of U.S. workers suffering from mental health issues. A stronger employee experience and a greater level of engagement are achieved when employees are supported through an actual wellness programme in the workplace.

4.9 Reduction in Absenteeism

Every worker must be absent from work once in a while but when those days are too frequently off, it becomes an issue of absence and is known as absenteeism. This could indicate that before the activity starts to have a negative impact on the company, some adjustments in employee engagement need to be made within the organisation and work environment.

References

1. https://www.quantumworkplace.com/future-of-work/what-is-employee-engagement-definition/. Retrieved May 09, 2024.
2. Rao, S., Chitranshi, J., & Punjabi, N. (2020). Role of artificial intelligence in employee engagement and retention. *Journal of Applied Management-Jidnyasa,* 42–60.

Chapter 5
Employee Engagement Models

Fostering employee engagement goes beyond just sustaining the business and maximising profits. It is also about ensuring that employees feel content and fulfilled. With the World Health Organisation officially recognising burnout as an "occupational phenomenon" three years ago, prioritising the well-being of employees as individuals who live outside of work has become increasingly crucial in reducing employee turnover and maintaining business success. An employee engagement model serves as a framework that delineates how a company can support its employees in feeling happy, content, supported, appreciated, respected and trusted in their workplace. It also acknowledges that employees are individuals with lives both within and beyond the workplace. Models of employee engagement extend beyond mere compensation or traditional benefits (although these are still important). They establish the foundation for the company's culture and guide the methods of leadership and management. There are six models of employee engagement, i.e., (i) the Zinger model, (ii) the Gallup model, (iii) the AON-Hewitt model, (iv) the Kahn model, (v) the Maslow model and (vi) the JD-R model (Fig. 5.1).

5.1 The Zinger Model

The model was created by psychologists and educators David Zinger; this model focuses on leveraging connections between people to achieve better results. It focuses on strategy and organisation, recognising employees' finite energy as a key factor in whether they attain the goal of genuine happiness. It also considers professional development to be equally important for serving customers. It places emphasis on strategy and organisation, acknowledging the limited energy of employees as a crucial factor in their pursuit of genuine happiness. Additionally, it recognises the significance of professional development in serving customers. The Zinger employee engagement model revolves around the acronym CARE, representing connection,

Fig. 5.1 Employee engagement models

authentic relationships, recognition and acknowledgement of employee effort, and continuous engagement. The Zinger model is supreme for workplaces that identify the effects of spiritual, emotional and mental force levels on proficient associations and efficiency.

5.2 The Gallup Model

The Gallup model is highly adaptable and effective for teams of any size and across different sectors. It is particularly valuable for remote, hybrid or distributed teams, as it enables leaders to gauge satisfaction using a small set of key questions in an employee engagement survey sent through a platform such as Leapsome. Gallup emphasised that all leaders and managers should take responsibility for enhancing employee engagement and not view it solely as an "HR issue." Through surveying thousands of US workers over the past 50 years, the organisation has gained insights that show that engagement revolves around providing employees with purpose, ensuring their development, assigning caring and supportive managers, conducting regular conversations and focusing on strengths.

5.3 The AON-Hewitt Model

The AON-Hewitt model places a strong emphasis on achieving positive business results and driving employee engagement. Engaged employees have a direct impact on both customer satisfaction and a company's profitability. This model of employee engagement integrates six fundamental components that operate in conjunction:

- Fundamental aspects include employee benefits, job security, workplace safety and maintaining a healthy work-life balance.
- Organisational policies and procedures encompassing communication, diversity, inclusion and infrastructure.
- Job responsibilities include fostering collaboration, granting empowerment and providing autonomy.
- Leadership, which encompasses both middle management and senior executives.
- Performance evaluation includes a structured career advancement framework, further career opportunities, rewards and acknowledgement.
- Brand—for example, corporate accountability and company status.

These components come together from the traits of Say, Stay and Strive, which are exhibited by employees who will speak positively about their company, wish to stay because they have a strong feeling of loyalty to their employer, and work hard to do their best work.

Larger businesses may find success with the AON-Hewitt approach due to its succinct Say, Stay and Strive language.

5.4 The Kahn Model

This employee engagement model was developed and named after organisational psychologist William Kahn. It places a strong emphasis on employees' perceptions of their strengths being reflected in their jobs. The Kahn model's overall goal is for every worker to feel secure enough to "bring their full self to work" and utilise their unique talents to produce excellent outcomes. This concept may be especially useful for businesses that wish to promote transparency, empathy, and compassion among their employees. According to Kahn, for employees to be fully and effectively engaged, they need to do the following:

- Feel that their work has meaning for them.
- Feel comfortable bringing their whole selves to work without fear of negative feedback.
- Feel that they are both physically and cognitively capable of doing so.

5.5 The Maslow Model

This model focuses on how employee experience in the workplace might be impacted by Maslow's well-known hierarchy of needs theory. According to Maslow's hierarchy, human needs are arranged in a pyramid, with each level needing to be met before moving on to the next. The order from lowest to highest levels is as follows:

(a) Food, drink, clothes, air, shelter, and rest;
(b) Security and safety;
(c) Love and a sense of belonging;
(d) Self-worth;
(e) Self-actualisation.

According to the Stress Management Society, the top four levels of the pyramid are the most persuasive in regard to applying Maslow's concept to employee engagement. For employees to be engaged, they must be satisfied at the very least at the basic level, which includes having enough money to cover their basic needs and getting adequate sleep. At the next tiers, managers can achieve significant increases in employee engagement. Because Maslow's original hierarchy is well-known and straightforward, it can be simpler to apply throughout the entire organisation and communicate to employees than other models.

5.6 The JD-R Model

The acronym JD-R refers to job demands-resources, and it characterises workplace stress and pressure as an imbalance between these two fundamental components:

- Job demands encompass any job, whether physical or psychological, that requires something from employees. This encompasses work pressure, social stress, exertion or the utilisation of skills.
- Job resources encompass any aspects that provide something to employees, such as growth opportunities, independence or clarity regarding their roles.

When a job is excessively demanding, it can lead to physical and mental exhaustion among employees, resulting in decreased performance, excessive strain, reduced engagement and health issues. With ample resources, employees are more likely to realise their potential, perform effectively, avoid burnout and feel supported. The model could be effective for challenging positions (such as high-stress or physically demanding jobs) or for organisations seeking to prioritise employee growth opportunities and additional support.

Chapter 6
Artificial Intelligence in Employee Engagement and Retention

Artificial intelligence (AI) is similar to a wizard that helps employees determine the mysteries of the workforce. The fact that engaged workers are the engine of any successful organisation is no secret. Employees are passionate souls who bring their game to work, fuel innovation and drive productivity up through the roof. AI can analyse vast amounts of employee data with speed and precision. From employee performance metrics and employee feedback to sentiment analysis and social interactions, AI can crunch the numbers and detect patterns that the human eye might miss, but it is not just about numbers and statistics, folks. AI has the power to understand the emotional side of employee engagement as well. By analysing employee feedback and sentiment in employee communications, AI can gauge overall mood, identify potential issues, and provide invaluable insights for improvement. It can be imagined that a crystal ball predicts which employees may be at risk of being disinclined to work for the company or leave it altogether. To maintain top talent loyalty and motivation, AI allows proactive action to be taken by delivering personalised solutions and support. However, AI can also help employee development programmes identify trends, skill gaps, and recommend training opportunities tailoring to individual needs. It is like having an AI-powered career coach, cheering employees towards growth and success. This is not the only thing that AI can do; it can help to design targeted employee development programmes, detect trends and skills gaps and recommend training opportunities that are specific to individual needs [1].

The main factor in any organisation's success is the involvement of employees. It is a measure of employees' commitment to their organisation. This is an important factor for any company because it has a major impact on its culture and productivity, management team and workers' working life balance. By providing employees with information about what they want and need from their employers, artificial intelligence can be used to support companies engaged in employee engagement [2]. AI can help in monitoring employees' mental wellbeing, behavioural patterns and engagement levels through real-time data collected through emails, chats, and facial recognition (facial expressions). By analysing historic and current data, identifying patterns, anticipating trends and providing personalised solutions using the prescriptive analytics platform, it will help to measure levels of involvement.

For instance, data-driven insights can assist in classifying employees who are unhappy and intend to leave the organisation. Artificial intelligence (AI) can quickly identify this group of disgruntled workers and assist HR departments in keeping them on board. Collaboration tools that include businesses use a variety of technologies to foster employee companionship. Numerous committed groups are being formed on the communicator in the company. Thus, working with colleagues who share similar interests or skill sets is beneficial. In turn, this supports the organisation's efforts to maintain motivated and engaged staff. The AI keeps an eye on all of this in real time. Learning that is targeted means that AI-enabled learning applications and platforms also inquire about the learner's desired level of knowledge and highlight any gaps. A recommended individualised learning path will be made based on the data gathered. This greatly personalises the material and ultimately sustains employee motivation through to the conclusion. Real-time input refers to gathering real-time engagement feedback, which is one of the best methods for determining exactly what is going through employees' minds, including what emotions they are displaying and what topics they are most interested in talking about at work. Businesses can obtain real-time employee input and pinpoint the precise areas where they need to focus their efforts by utilising AI-powered feedback tools. Chatbots are AI-driven communication platforms, and these automated conversations are so dynamic that they can involve conversations with employees about what they need. Feelings are communicated while completing the task at hand [3]. Within the company, the chatbot serves as a reminder to approve timesheets, acknowledge team member performance, and provide assistance with policies, contract letters, utility connection address letters, and critical contact information. Employers can utilise these data to adjust their management approach and organisational culture to increase employee engagement (Fig. 6.1). Nevertheless, the advantages of AI go beyond higher levels of engagement. AI can be used to provide or receive input from workers regarding how to improve their work-life balance. At that point, organisations might adapt or transform themselves as needed [4].

Businesses are employing analytics tools to go far deeper into their data to increase efficiency, gain a competitive advantage, and improve their bottom lines in terms of both real and intangible assets in today's corporate environment. For this reason, businesses are eager to include and use artificial intelligence (AI) to obtain outcomes that are quicker and more accurate. Artificial intelligence has also made a difference in enterprises by eliminating monotonous tasks. Because AI plays many different roles in business, from enhancing employee and customer relations to finding patterns in massive volumes of data to automating repetitive tasks and processes, every industry in the world is discreetly planning to use it. The objective is to seize the chance that artificial intelligence presents for worker productivity and success [5].

Since AI can analyse large amounts of data and produce trend directions and useful recommendations, implementing it might be a crucial tool for any business seeking quantitative assistance in its decision-making processes. AI facilitates the setting and

Fig. 6.1 Artificial intelligence in employee engagement

attainment of better objectives and outcomes by managers, staff, and companies. By offering insights into AI, which will usher in a new era in the industry, this study will highlight the impact of using AI to enhance employee behaviour and work outcomes [6]. It is based on the use of artificial intelligence and its impact on HRM because of technological advancements in the IT landscape.

Talent retention is considered the greatest risk in industry worldwide. Businesses are always attempting to devise new plans and techniques for keeping their staff members. The human resource specialist is largely responsible for this task [6]. In this cutthroat market, keeping personnel in the company presents a problem for any HR executive. HR uses a variety of tactics to address this issue within the company. As a result, AI can be extremely helpful in career advancement planning by monitoring and evaluating worker performance to identify depressive and stagnant patterns. This will assist managers and strategists in organising the launch of fresh learning and development (L&D) initiatives and locating fresh chances for expansion.

High attrition can be detrimental to an organisation in a competitive setting, particularly if it affects its best employees. Top performers are typically difficult to replace, and when teams lose people who have a critical understanding of a product or process, productivity is negatively impacted [7]. Creating a retention plan that makes use of cutting-edge technology. There are few strategies for applying AI to increase staff retention.

6.1 Career Planning

In reality, the majority of businesses do not know what kind of workers they want in ten years. This is typically the case for career advancement. Therefore, giving each worker the opportunity to sketch out a precise path to their next position within the company can help to inspire and involve them [8]. One of the main factors influencing an employee's decision to stay at a company is the possibility of career advancement. Artificial intelligence (AI) can be a valuable tool in career progression planning by tracking employee performance and identifying stagnation. This will assist managers in finding new growth possibilities and organising the launch of new learning and development (L&D) programmes.

6.2 Identifying Opportunities for Growth

AI can be used to measure employee engagement and quality of work performance by providing managers timely reminders about when an employee might become ready for a new challenge or if he or she has grown too much involved in his or her role. To make recommendations on the next steps for training their staff, managers may also use AI by means of performance management tools. Furthermore, AI assistance can help employers develop a high-touch approach to the development of employees and maintain workers' potential at risk for turnover.

6.3 Better Work-Life Balance

Work-life balance identifies the equilibrium between an individual's priorities at work and their priorities in other aspects of life. One of the most important factors for employee satisfaction is a balance between work and life. As more employees demand greater control over their working time and schedules, AI can help people give them the autonomy they want while keeping work demands in mind.

6.4 Achieve Equitable Compensation

As businesses strive to remain competitive in today's economy and as there is a greater focus on equal pay for equal work, the question of how to best compensate employees becomes more important. The ability of artificial intelligence to analyse large amounts of data and identify patterns will allow employers to create more tailored and equitable compensation packages for their employees with the help of AI. To optimise remuneration and benefits, these solutions can assess market conditions, employee performance and business results. To achieve better results, companies may also set more clearly defined targets, monitor the progress of their employees and advise on new methods. In the end, AI can help employers create a more conducive working environment for their employees.

6.5 Improving Productivity

To increase an organisation's productivity, better retention is beneficial. However, improved long-term retention can be achieved through the use of AI even during the recruitment process. People who are a lot of fitters for the job will stay longer because they are making good money. They are being recognised. They love what they do, and the natural result is a higher retention rate. There may also be a great deal of influence from small changes. For example, an employee who shows up late on a periodic basis when the work is not done. Teams need to wait until the employee starts. Productivity is reduced. The surprising way in which AI can help is to identify patterns of employee tardiness. Ultimately, AI is not about being cruel or indifferent. To create an environment where employees are encouraged to do their best work, data are applied.

AI will not be able to completely replace human judgments in the near future, but these systems can help people make more informed choices and are likely to have stronger support. If the system is capable of properly learning and adapting behaviour, it will be much easier to run a variety of scenarios and test cases. They will be able to fill their company with people who are suitable for other members of their team, thus ensuring a more harmonious working environment [9]. The tools provided by inhuman AI can be used by human resource managers, allowing them to take advantage of their imagination and creativity, two qualities that make them more human. Combining big data with human touch is the key to the effective use of AI to store data. Algorithms are just as smart as the people writing them. A human resources team that knows how to balance the use of technology with subtle nuances in working with people should be ensured by any company considering the use of artificial intelligence for better retention and development of its staff.

References

1. Rao, S., Chitranshi, J., & Punjabi, N. (2020). Role of artificial intelligence in employee engagement and retention. *Journal of Applied Management-Jidnyasa*, 42–60.
2. Rožman, M., Oreški, D., & Tominc, P. (2022). Integrating artificial intelligence into a talent management model to increase the work engagement and performance of enterprises. *Frontiers in Psychology, 13*, 1014434.
3. Hughes, C., Robert, L., Frady, K., & Arroyos, A. (2019). Artificial intelligence, employee engagement, fairness, and job outcomes. In *Managing technology and middle- and low-skilled employees* (pp. 61–68). Emerald Publishing Limited.

4. https://medium.com/@manjunath.dharmatti/ai-in-employee-engagement-973da848ad96. Retrieved May 10, 2024.
5. Lourens, M., Krishna, S. H., Singh, A., Dey, S. K., Pant, B., & Sharma, T. (2022, December). Role of artificial intelligence in formative employee engagement. In *2022 11th International Conference on System Modelling & Advancement in Research Trends (SMART)* (pp. 936–941). IEEE.
6. Burnett, J. R., & Lisk, T. C. (2021). The future of employee engagement: Real-time monitoring and digital tools for engaging a workforce. In *International perspectives on employee engagement* (pp. 117–128). Routledge.
7. Rodgers, W., Murray, J. M., Stefanidis, A., Degbey, W. Y., & Tarba, S. Y. (2023). An artificial intelligence algorithmic approach to ethical decision-making in human resource management processes. *Human Resource Management Review, 33*(1), Article 100925.
8. Das, S., Chakraborty, S., Sajjan, G., Majumder, S., Dey, N., & Tavares, J. M. R. (2022, December). Explainable AI for predictive analytics on employee attrition. In *International Conference on Soft Computing and Its Engineering Applications* (pp. 147–157). Springer Nature Switzerland.
9. Dutta, D., Mishra, S. K., & Tyagi, D. (2023). Augmented employee voice and employee engagement using artificial intelligence-enabled chatbots: A field study. *The International Journal of Human Resource Management, 34*(12), 2451–2480.

Chapter 7
Leveraging AI for Employee Engagement

AI has become a game changer in the modern workplace in regard to engagement with employees. Vast amounts of data can be analysed using AI technologies such as machine learning algorithms and natural language processing, which can provide valuable information. This allows organisations to gain a better understanding of the behaviour, preferences and needs of their staff. Using artificial intelligence, organisations can identify patterns, anticipate outcomes and adapt their engagement strategies according to them [1]. The impact of AI on employees is multifaceted, ushering in transformative changes across various aspects of the professional landscape. First, AI has substantially altered the way natural language processing is carried out at work through automated tasks that allow workers to focus their attention on more complicated and exciting aspects of their jobs [2]. As employees undertake higher value tasks, this transition improves job satisfaction and gives them a chance to improve their skills.

In addition, AI-enabled analytics in performance evaluations have contributed to a proper and unbiased assessment of employees' contributions. AI helps to create a culture of merit, which ensures that recognition and promotion opportunities are based on individual achievements rather than subjective judgments, by removing potential biases. On the other hand, as advances in AI-powered employee engagement technology emerge, concerns about job displacement have become apparent. Automation may lead to obsolete roles, requiring redeployment efforts to ensure that workers remain relevant as the labour market evolves [3]. However, the positive impact of AI on employee welfare is not limited. AI can analyse a variety of indicators, such as workload and stress levels, to provide organisations with insight into the implementation of measures designed to promote healthy working life balance [3].

© The Author(s), under exclusive license to Springer Nature Singapore Pte Ltd. 2025
S. Majumder and B. Misra, *Analysing Trends and Patterns in Employee Engagement Through AI*, SpringerBriefs in Computational Intelligence,
https://doi.org/10.1007/978-981-96-4496-4_7

7.1 AI for Employee Recognition and Rewards

Recognition and rewards play important roles in fostering employee engagement. With AI, organisations can take employee recognition and rewards to a whole new level to make the experience more personal and relevant. By providing valuable insight and feedback, AI-powered recognition systems are revolutionising the evaluation of performance. Through AI's ability to analyse employee data, areas for improvement and recommend custom development plans can be identified and performance evaluations can be improved.

7.1.1 Key Steps to Develop Effective Automated Recognition Systems

Based on preset criteria, automated rewards systems software honours and rewards staff members for their diligence, successes, actions and milestones. Immediate, focused rewards are intended to enhance the entire work experience. The purpose of automated rewards and recognition programmes is to provide employees with material prizes—such as money or gift cards—for acts that support company objectives [4]. This kind of programme is simple to administer, set up and monitor for employee development. Automated recognition and reward programs can improve worker morale, performance and engagement when properly implemented [5]. The main steps involved here are as follows:

(a) **Clear Vision of the Programme Goals**

Beyond raising staff morale, they ought to be well aware of the business advantages anticipated from the initiative. They also need to be aware of the behaviours and values that they hope to instil in staff members through the recognition programme [6]. Therefore, laying out the programme's objectives clearly and informing staff about them can assist in building a solid foundation for success.

(b) **Seek Input from a Cross-Section of Employees**

Their recommendations regarding the procedures, eligibility requirements, and categories of acknowledgement should be appropriately included. As a result, this fosters trust among staff members and helps them feel like essential members of the programme. Considering their ideas and viewpoints also increases the system's credibility [7].

(c) **Set Up a Rewards and Recognition Steering Committee**

It ensures fairness and transparency with proper representation from all stakeholders. The committee can also handle the nitty–gritty, including organising regular reward functions and deciding on the best rewards [8].

(d) **Development of the Nomination and Selection Guidelines**

7.2 Gamification and AI-Powered Incentives

To guarantee that there are sufficient possibilities for every employee to appreciate, HR must collaborate closely with the committee. To provide employees at all organisational levels with equitable opportunity, rewards might fall into several categories [9].

(e) **Announce and Launch the Programme**

Companies ought to send each employee an email outlining the different components of the programme. Furthermore, to encourage employee interest, confidence and programme engagement, the incentive committee should respond to any questions [10, 11].

7.1.2 Advantages of Automated Systems

(a) Greater accuracy—they track rewards and recognition more quickly and reliably than manual methods.
(b) Automation can guarantee that workers are paid on schedule.
(c) Monitoring staff performance is made simpler by automated technologies that allow managers to keep an eye on developments and provide regular feedback.
(d) Scalability is provided by automation; businesses may swiftly scale up or down operations without having to invest more personnel or resources.

7.1.3 The Challenges of Automated Systems Include

(a) It may take some time for everyone to become accustomed to automation because some staff members may find the new technology intimidating.
(b) Technical issues may arise during the transition period or afterwards, slowing down operations and disrupting workflows.
(c) It may be expensive to implement an automated system, particularly if the current IT infrastructure is not on par with the required capabilities.

7.2 Gamification and AI-Powered Incentives

Gamification is a powerful tool for motivating employees, and artificial intelligence enables organisations to develop effective gamified solutions. With the aid of artificial intelligence, organisations can design incentive schemes for each employee according to their preferences and interests. To identify major performance indicators and ensure the alignment of rewards with staff achievements, AI algorithms can interpret data [12].

Gamification is a concept that aims to increase engagement and commitment by adding fun and excitement to routine tasks or interactions. In the business world, this approach can be used for performance management, customer engagement, employee training, and product development [13]. For example, a sales team could use a points-based system where members receive rewards for hitting targets, making a boring task more competitive and exciting.

Psychological concepts such as the need for approval, the thrill of competition and the need for recognition are also leveraged by gamification. When these components are included in corporate procedures, motivation and output can be greatly increased. This strategy helps workers or users who are immediately impacted as well as wider organisational objectives, including greater sales, improved customer service and increased efficiency [14].

The synergistic combination of AI with gamification in business environments can greatly increase the efficacy of both technologies. AI improves gamification through challenge optimisation, experience personalisation and actionable insights.

7.2.1 Personalisation

Personalisation is a major area in which AI shines. AI may assess user data in gamified systems to customise interactions, challenges and rewards according to user preferences and historical conduct. By tailoring the gamification experience to each individual user, this personalised method boosts motivation and involvement.

7.2.2 Predictive Insights

AI can also use predictive analytics to maximise the gamification process. AI is able to forecast which game components will work best for certain user segments by evaluating patterns and results. For instance, in a work environment, AI may identify which challenges or rewards most motivate particular employee groups, resulting in more successful engagement tactics.

7.2.3 Testing and Improvement

Product managers can improve user experience and optimise tactics by using AI for A/B testing and insights into gamification aspects.

7.2.4 Dynamic Content

AI-generated in-game content that is generated in response to user preferences and progress, saving time and effort when creating new content while maintaining the game's originality.

7.2.5 Fraud Prevention

AI's function in identifying and stopping dishonest or fraudulent activity to maintain an enjoyable and equitable gaming environment.

7.2.6 Comments Analysis

By effectively analysing player reviews and comments with the use of AI-powered sentiment analysis, problems can be quickly identified, and chances for development can be quickly realised.

7.2.7 AI-Based Player Segmentation Targets Specific Gamification Tactics and Rewards to Increase Player Engagement

Furthermore, AI can use gamification data to extract insightful information. AI is able to recognise patterns, achievements and areas for development through the analysis of user interactions and results. Gamification strategies can be continuously improved using these data, guaranteeing their efficacy and alignment with organisational objectives [15].

7.3 AI-Enabled Performance Management

In an objective assessment of performance, artificial intelligence plays a critical role. Organisations can use artificial intelligence algorithms to assess performance on the basis of quantitative metrics and eliminate biases and subjective assessments. The ability of AI to analyse employee data allows for continuous feedback and improvement, which enables organisations to provide early feedback and propose development options. Moreover, organisations are empowered to predict trends in performance and take progressive action on the growth of staff by way of Predictive Analytics in Performance Management [16].

The importance of AI in today's data-driven environment cannot be emphasised. Artificial intelligence possesses the ability to handle and analyse enormous volumes of data at speeds that are unattainable for humans. This is where performance management has started to notice AI's importance. Artificial intelligence (AI) can quickly pass through the massive volumes of data that businesses analyse about employee performance, uncovering trends, patterns, and insights that might otherwise go unnoticed [17]. When artificial intelligence (AI) is included in performance management procedures, decision-making becomes more accurate, objective and efficient.

7.3.1 Automated Performance Review

The greatest benefit of AI is that performance reviews can be automated. By employing AI algorithms for data collection, analysis, and integration, AI automates methods for evaluating an individual's performance. It supports team leaders' ability to conduct employee performance reviews and concentrate on accurate information.

7.3.2 Seamless Data Collection

The AI system is not restricted to particular or generalised data based on fundamental characterisation; rather, it is made to gather data from a variety of sources [18]. It examines a variety of factors and characteristics, including individual performance, human nature, career advancement, role in corporate operations, customer experience, emotional instability or intensity in employees' stability based on activity trends, feedback from other employees, etc. [19]. Artificial intelligence is seen as significant since it essentially offers a 360° study of performance reviews.

7.3.3 Reduced Bias

AI can provide a more objective assessment of employee performance by removing some of the human biases that may exist in traditional performance appraisals. AI measures performance using indicators such as work performance, analytics from team leaders and other non-perception-based measurements. Hence, artificial intelligence in performance management systems can assist individuals in ignoring these biased areas and guaranteeing equality [20]. The use of artificial intelligence for performance is believed to eliminate bias based on factors such as age, race, gender, ethnicity and nationality and provides everyone with equal opportunities.

7.3.4 Real-Time Data Analysis

With AI-driven performance management solutions, management can quickly and easily analyse data in real time, facilitating prompt and well-informed decision-making. Performance monitoring is no longer limited to yearly evaluations. Real-time updates on employee progress can be obtained via AI-powered systems, enabling prompt goal adjustments and interventions. This is crucial since inconsistent feedback might affect an employee's overall performance. Thus, prompt feedback helps identify areas for improvement right away and increase productivity in the workplace.

7.4 Enhancing Employee Communication and Collaboration with AI

The basis of a successful workplace is communication and cooperation. With AI, it is possible to streamline internal communications, automate routine tasks and improve the decision-making process. Since misunderstandings occur frequently at work, effective communication techniques are crucial in any setting. Artificial intelligence has the capability to automate repetitive responses, allowing employees to quickly locate answers to their questions [21]. It frees CEOs, HR specialists and company leaders from the burden of questions and answers so that they can concentrate on higher-value work. Simple contact with chatbots and other AI technologies provides further assistance for organisational development and expansion.

The field of AI has completely changed how humans interact. Artificial intelligence has facilitated faster, easier, and more efficient communication through developments in machine learning and natural language processing. With their ability to provide prompt, individualised support, these AI-powered chatbots and virtual assistants have completely changed how people communicate in the workplace [22]. They may help with employee inquiries, walk them through HR procedures, and give them rapid access to pertinent data. They are available 24/7. This immediate assistance creates a sense of efficiency and trust by streamlining HR procedures and enabling staff members to address problems immediately. It is also mutually beneficial because expediting these procedures frees up the communications team's time for more strategic or innovative projects [23].

Personalising one's internal communications based on the tastes, requirements, and conduct of staff members is another way that AI may enhance internal communication [24]. AI, for instance, may assist in audience segmentation, message customisation and delivery via the most efficient channels and formats. AI may also assist in developing more interactive and engaging communication. To strengthen bonds with staff, AI may also assist in analysing the mood, tone and emotion of communication and making necessary adjustments.

7.4.1 Translating and Customisation

AI can help multinational corporations overcome language barriers by offering real-time translation and localisation services and facilitating cross-border collaboration.

7.4.2 Improving Cooperation

AI can also improve internal communication by helping staff members collaborate, particularly when they are spread out across different departments, time zones or regions. To enhance employees' learning and growth, AI can also help to link them with appropriate materials, knowledge and professionals. AI can assist in cultivating an innovative culture by facilitating the exchange of ideas, feedback and best practices among staff members and by honouring and rewarding their efforts.

7.4.3 Personalisation

AI has the ability to modify communications according to a person's preferences and previous exchanges. This degree of customisation can improve the relevance and engagement of communications. Additionally, AI can support the development of more dynamic and engaging communication through the use of gamification, polls, quizzes and videos to boost participation and feedback.

7.4.4 Streamlining Regular Conversations

Regular interactions, such as scheduling, reminders and updates, can be automated by AI. This saves time and frees staff members from working on more important projects. Furthermore, AI-powered chatbots can answer a wide range of simple questions, freeing up human resources for trickier problems.

7.4.5 Commercial Interactions

It can be applied to enhance the voice output of a chatbot or facilitate the process of looking up specific information. AI is also being increasingly utilised in social networks, for instance, to display tailored advertisements. AI is applicable here not only for content production but also for communication measure analysis and targeting.

7.5 Intelligent Virtual Assistants

By providing real-time support and assistance to employees, AI chatbots such as Albus are revolutionising human communication. The communication channels are simplified, the response times are reduced, and the efficiency is increased by these virtual assistants. In addition, AI chatbots may be able to perform routine tasks and free up valuable time for employees focused on higher-value jobs. It is like having a virtual teammate who does not sleep, which keeps the collaboration spirit alive all day [25]. This helps to improve productivity and increase employee satisfaction by lessening tedious tasks.

Intelligent virtual assistants, or IVAs, are highly advanced software programmes that have the power to completely change how contemporary companies provide customer care. HR departments mainly rely on in-person commercial transactions [26]. The way HR operates has now altered due to AI chatbots and virtual assistants, which make it simple for HR professionals to automate routine employee interactions. AI-powered personal assistants, such as Google Assistant, Alexa and Siri, have gained much popularity in recent years. NLP and machine learning methods are used by these virtual assistants to understand and respond to user orders and questions. They are capable of carrying out a great deal of duties, such as playing music, controlling smart home appliances, and sending and setting reminders. Conversely, IVA is a result of advancements in machine learning (ML) and artificial intelligence (AI). Comprehending these fundamentals aids in deciphering the intricacies of IVAs [27].

7.5.1 *Natural Language Processing (NLP)*

An intelligent virtual assistant's capacity to comprehend and analyse human language is essential to its efficacy. IVAs can grasp context and user inquiries through natural language processing (NLP). Interactions with an IVA are spontaneous and casual because of NLP.

7.5.2 *Responsive Learning and Predictive AI*

Predictive AI produces predictions and recommendations based on past data. This is what gives IVAs the ability to predict user needs. An IVA can pick up knowledge from user input and change over time, gratitude to predictive AI and adaptive learning. It can predict what one person could ask next in addition to understanding what the person has asked. Interaction with an IVA will improve people's ability to help them.

7.5.3 Generative AI

Although generative AI is used less frequently in IVAs, it is nevertheless useful for producing answers to challenging questions or responses that sound more genuine. IVAs can provide more dynamic user experiences by responding with more than simply generic or templated responses.

7.6 Augmenting Decision-Making with AI

AI gives organisations the power to make data-driven, informed decisions. AI-based decision support systems analyse large amounts of data to gain valuable insights and enable organisations to make the right choices. By integrating AI algorithms, organisations can enhance employee engagement by empowering employees with the tools and information needed to participate in decision-making processes. This fosters a sense of ownership and inclusion among employees.

7.7 Improving Employee Development and Learning with AI

The growth and engagement of employees requires continuous learning and development of skills. To improve the development of staff and promote a culture of education, AI is offering innovation solutions. Using machine learning and natural language processing techniques, artificial intelligence provides an excellent learning experience for each employee. This technology allows workers to take responsibility for their learning journeys. It automates mundane tasks and accelerates training programmes at lightning speeds to support research and development teams.

7.8 Virtual Reality and AI for Immersive Training

Immersive and meaningful training experiences can be offered through VRVR, together with artificial intelligence. Virtual reality enhanced with artificial intelligence allows employees to practice and improve their skills in a safe and controlled environment by simulating real-world scenarios. AI algorithms can adapt the training experience based on employee performance and learning preferences, ensuring an engaging and effective learning experience.

7.9 Measure and Manage Employee Sentiment

In the past, to improve employee morale, employee engagement has been considered to be a single incentive or initiative. This strategy was largely reactive to changes within the organisation (high turnover rates, low productivity, etc.). AI changes that leverage the power of data mining and machine learning and AI engagement solutions help HR leaders analyse and predict employee needs and behaviour.

7.10 Improving Personalisation

Traditional engagement initiatives took a one-size-fits-all approach to building engagement—no surprising is why such initiatives did not yield tangible results in the long run. Today's workers are tech-savvy digital natives, especially in the younger generation. They expect the same level of personalisation at work as they would expect from their customers.

7.11 Obtaining Support from Leadership

Getting support from leadership is essential for the successful implementation of AI projects. Company owner should discuss how to present a strong business case, show possible ROIs, and draw attention to the advantages that AI can offer in terms of competitiveness. Organisations can gain the required leadership support by clearly articulating the value and long-term advantages of AI.

References

1. Hughes, C., Robert, L., Frady, K., & Arroyos, A. (2019). Artificial intelligence, employee engagement, fairness, and job outcomes. In *Managing technology and middle- and low-skilled employees* (pp. 61–68). Emerald Publishing Limited.
2. Lourens, M., Krishna, S. H., Singh, A., Dey, S. K., Pant, B., & Sharma, T. (2022, December). Role of artificial intelligence in formative employee engagement. In *2022 11th International Conference on System Modelling & Advancement in Research Trends (SMART)* (pp. 936–941). IEEE.
3. Burnett, J. R., & Lisk, T. C. (2021). The future of employee engagement: Real-time monitoring and digital tools for engaging a workforce. In *International perspectives on employee engagement* (pp. 117–128). Routledge.
4. Sobolu, R., Stanca, L., & Bodog, S. A. (2023). Automated recognition systems: Theoretical and practical implementation of active learning for extracting knowledge in image-based transfer learning of living organisms. *International Journal of Computers Communications & Control, 18*(6).

5. Gadekar, M. A., Chandak, M. K., Budhalani, B. S., Chawale, S. S., Des's, C. O. E. T., & Rly, D. Automated employee face recognition system.
6. Boyd, K. L., & Andalibi, N. (2023). Automated emotion recognition in the workplace: How proposed technologies reveal potential futures of work. *Proceedings of the ACM on Human-Computer Interaction, 7*(CSCW1), 1–37.
7. Robert, L. P., Pierce, C., Marquis, L., Kim, S., & Alahmad, R. (2020). Designing fair AI for managing employees in organisations: A review, critique, and design agenda. *Human-Computer Interaction, 35*(5–6), 545–575.
8. Schweyer, A. (2018). Predictive analytics and artificial intelligence in people management. *Incentive Research Foundation*, 1–18.
9. Ramachandran, K. K., Mary, A. A. S., Hawladar, S., Asokk, D., Bhaskar, B., & Pitroda, J. R. (2022). Machine learning and role of artificial intelligence in optimising work performance and employee behavior. *Materials Today: Proceedings, 51*, 2327–2331.
10. Hooi, L. W., & Chan, A. J. (2022). Innovative culture and rewards-recognition matter in linking transformational leadership to workplace digitalisation? *Leadership & Organisation Development Journal, 43*(7), 1063–1079.
11. Clark, D. (2020). *Artificial intelligence for learning: How to use AI to support employee development*. Kogan Page Publishers.
12. Hinton, G. (2021). Empowering learning: AI-driven adaptive gamification and personalised learning in the digital world.
13. Khan, S. (2023). Elevating enterprise agility: Gamification and AI-powered ERP for dynamic performance enhancement. *Social Sciences Spectrum, 2*(1), 139–145.
14. Juli, M. (2024). *Gamified AI-ERP fusion: Maximising business performance through innovative synergy* (No. 12748). EasyChair.
15. Qureshi, H. (2023). Boosting enterprise performance: Leveraging gamification and text analytics with ERP and AI. *Social Sciences Spectrum, 2*(2), 154–159.
16. Majumder, S., & Dey, N. (2022). *AI-empowered knowledge management*. Springer.
17. Sundaresan, S., & Zhang, Z. (2022). AI-enabled knowledge sharing and learning: Redesigning roles and processes. *International Journal of Organisational Analysis, 30*(4), 983–999.
18. Hamilton, R. H., & Davison, H. K. (2022). Legal and ethical challenges for HR in machine learning. *Employee Responsibilities and Rights Journal, 34*(1), 19–39.
19. Sundarapandiyan Natarajan, D. K. S., Subbaiah, B., Dhinakaran, D. P., Kumar, J. R., & Rajalakshmi, M. (2024). AI-powered strategies for talent management optimisation. *Journal of Informatics Education and Research, 4*(2).
20. Dhanalakshmi, R., Cherukuri, D. M., Ambashankar, A., Sivaraman, A., & Sood, K. (2023). Smart analytics and AI for managing modern performance management systems. In *Smart analytics, artificial intelligence and sustainable performance management in a global digitalised economy* (Vol. 110, pp. 243–263). Emerald Publishing Limited.
21. Chowdhury, S., Budhwar, P., Dey, P. K., Joel-Edgar, S., & Abadie, A. (2022). AI-employee collaboration and business performance: Integrating knowledge-based view, sociotechnical systems and organisational socialisation framework. *Journal of Business Research, 144*, 31–49.
22. Arslan, A., Cooper, C., Khan, Z., Golgeci, I., & Ali, I. (2022). Artificial intelligence and human workers interaction at team level: A conceptual assessment of the challenges and potential HRM strategies. *International Journal of Manpower, 43*(1), 75–88.
23. Pratt, M., Boudhane, M., Taskin, N., & Cakula, S. (2021). Use of AI for improving employee motivation and satisfaction. In *Educating Engineers for Future Industrial Revolutions: Proceedings of the 23rd International Conference on Interactive Collaborative Learning (ICL2020)* (Vol. 223, pp. 289–299). Springer International Publishing.
24. Turner, P., & Turner, P. (2020). Employee engagement and the employee experience. In *Employee engagement in contemporary organisations: Maintaining high productivity and sustained competitiveness* (pp. 1–26).
25. Ha, Q. A., Chen, J. V., Uy, H. U., & Capistrano, E. P. (2021). Exploring the privacy concerns in using intelligent virtual assistants under perspectives of information sensitivity and anthropomorphism. *International Journal of Human–Computer Interaction, 37*(6), 512–527.

26. Bolton, T., Dargahi, T., Belguith, S., Al-Rakhami, M. S., & Sodhro, A. H. (2021). On the security and privacy challenges of virtual assistants. *Sensors, 21*(7), 2312.
27. Beaver, I., & Mueen, A. (2020, April). Automated conversation review to surface virtual assistant misunderstandings: Reducing cost and increasing privacy. In *Proceedings of the AAAI Conference on Artificial Intelligence* (Vol. 34, No. 08, pp. 13140–13147).

Chapter 8
AI-Powered Analytics Uncover Hidden Patterns in Employee Engagement Data

It is possible to wonder whether artificial intelligence works its magic to reveal such unknown information. A powerful set of tools is brought to the table by AI, as discussed below.

8.1 Natural Language Processing (NLP)

The ability of AI to understand and interpret human language is enabled by its natural language processing skills [1]. It will help to understand the experiences and emotions of workers more deeply by analysing anonymised survey responses, identifying employees' opinions and identifying key themes. NLP lies at the core of these AI systems [2]. NLP enables computers to comprehend, analyse and work with human language in a way that is practical and meaningful for certain purposes [3]. NLP algorithms are capable of sorting complex data, extracting pertinent details and even picking up subtleties in languages and phrases in regard to resuming screening.

In the context of HR, natural language processing (NLP) refers to the analysis and conversion of unstructured HR data into insightful knowledge. This is how it operates:

(a) Gathering of data: Gathering the HR data that must be analysed is the first stage. Resumes, job descriptions, performance evaluations, employee comments and more can all fall under this category.
(b) Preprocessing and data cleansing: The data need to be cleaned and preprocessed before they can be examined. NLP approaches can standardise the data, find and fix spelling and grammar mistakes, and eliminate unnecessary information.
(c) Analysis of text: The text data must next be analysed to draw insightful conclusions. This may entail extracting themes, entities, attitudes and other information from the text using natural language processing (NLP) methods.

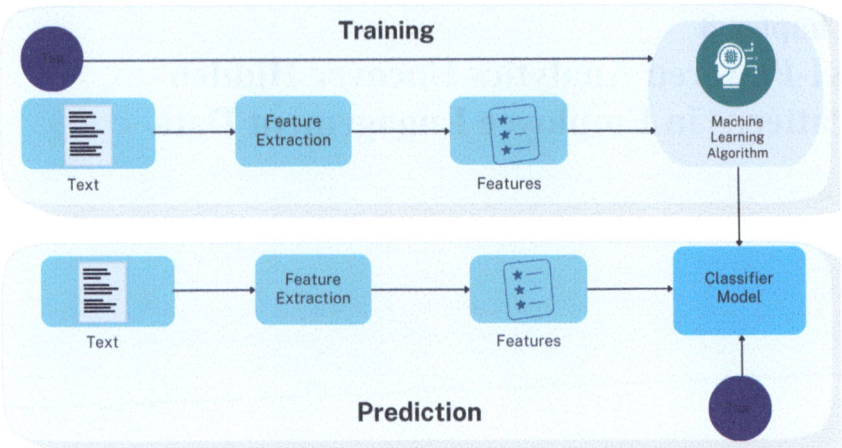

Fig. 8.1 NLP in data analysis and prediction

(d) Illustration: After the insights have been retrieved, dashboards, graphs, charts and other tools can be used to visualise the data. HR specialists will find it simpler to comprehend and analyse the data as a result.

(e) Making choice: Making data-driven decisions about HR practices and policies is possible with these insights. For instance, performance review data can be used to identify high-potential employees and provide them with development opportunities, and employee feedback analysis can be utilised to pinpoint areas where the workplace culture needs to improve.

In the end, NLP in HR converts unstructured HR data into insightful knowledge that can help with decision-making and enhance productivity at work. By combining NLP techniques [4] with HR knowledge, organisations may improve employee understanding and foster a successful work environment. The training and prediction process is illustrated in Fig. 8.1.

8.1.1 NLP in Sentiment Analysis

In essence, natural language generation is a branch of artificial intelligence (AI) that uses methods such as text analytics to help robots comprehend human language and provides a potent method for automatically analysing vast volumes of unstructured text data in people analytics. Sentiment analysis has been demonstrated to be effective in determining an organisation's skill set for appropriate workforce planning, and it may also be utilised to offer a more in-depth understanding of employee mood. After these scores are totalled, data visualisation dashboards, charts, or graphs are used to visually display the results to HR managers, corporate executives, and employee managers [5]. Business executives can enhance corporate culture and staff engagement by visualising employee emotion. With an emphasis on improving employee experience, employees can also use data to enhance their performance management procedures.

The goal of i-Pulse commitment presented in [6] seems to be to advance employee engagement efforts to a new degree. Through natural language text analysis, this AI-enabled technology offers deep insights into employee sentiments and assists logistics leaders in comprehending the worries of their workforce. Additionally, it gathers practical knowledge to pinpoint the root reason for disengagement within a company and develop innovative solutions to eradicate it. Insights into increasing organisational value to obtain a competitive edge in the logistics industry are provided by this study.

8.1.2 NLP in Employee Recruitment

Human resources departments and recruitment platforms can gain from meaningful matching through the detection of abilities, professional backgrounds and experiences indicated in a CV, professional profile or position by processing internal and external data. Businesses are unable to properly analyse and obtain insights from an excessive amount of unstructured data. Intelligent document analysis analyses material, extracts meaning and supports process automation and decision-making through the use of AI techniques such as NLP, entity extraction, semantic comprehension and machine learning. Better risk management, compliance and internal operational efficiencies are the ultimate results [7].

The majority of candidate search and match solutions are restricted to simple resume parsing and matching job advertisement metadata. With an advanced filter and a multicriteria search, NLP can personalise a search for applicants or job offers. NLP can also map internal skills to identify abilities or training needs [8]. Nevertheless, the bidirectional ideal solution goes beyond a straightforward keyword match and performs theme analysis, shortlisting the top candidates and positions and monitoring both. To provide statistical and linguistic capabilities for comprehending application profiles and finding the best candidates, it integrates advanced search algorithms, analytics, NLP and ML.

8.1.3 NLP in HR Chatbots

Timely access to appropriate information is crucial for the day-to-day operations of a company. Employees can easily locate the information they require about the company's structure, internal policies, and operational guidelines with the aid of an HR chatbot [9]. It offers a centralised point of access to business information and relieves administrative teams of repetitive questions when incorporated into a corporate chat system such as Slack.

8.1.4 NLP in Employee Engagement

Maintaining prospects' and workers' interest in their employment is a challenge for many organisations. However, NLP algorithms use text analytics to assess comments, create prediction models, and provide insight into the opinions and personal data of employees. Based on these automated data, HR may develop an employee engagement plan that successfully addresses their workplace issues.

An AI-engagement tool automatically evaluates thousands of employees' reviews or feedback, enabling the company to make necessary adjustments to its policies and maintain an advantage over its competitors. Many people, including employees and consumers, are engaged on the Ramesh platform [10]. As a result, one can utilise these NLP platforms to provide talent management insight into the mindsets and levels of satisfaction of the prospects.

8.2 Machine Learning

Natural language processing skills enable AI to understand and read human languages. Analysing anonymised responses to the surveys, identifying employees' opinions and identifying key themes will help us understand more about workers' experiences and feelings [11]. AI and ML tools are cutting-edge, contemporary technologies. They handle complex problems using coding programmes, analytics, and algorithms that are exactly the same as human logic and reasoning. HRM simplifies functions related to performance management, evaluation, communication and recognition. AI and machine intelligence boost employee engagement in modern firms in a variety of ways. These include gaining knowledge from performance information, automating routine work, developing individualised opportunities for learning and growth and enhancing diversity in the recruiting process to eliminate bias. Allowing artificial intelligence and machine learning to improve overall work experience allows the human workforce to concentrate more on higher-level activities. This indicates that workers are happier in their jobs.

Data collection, storage and analysis are subject to strict limitations due to compliance requirements and privacy restrictions, particularly in highly regulated areas such as healthcare and finance [12]. Thus, to protect sensitive employee data and guarantee legal compliance, businesses in these industries need to manage regulatory complications and establish strong DGFs. It is essential to analyse engagement data in businesses with high staff turnover rates, including retail and hospitality, to pinpoint attrition factors and put retention measures into place [13]. To improve

8.2 Machine Learning

employee satisfaction and loyalty, HR analysts in these industries must concentrate on preemptive risk assessment, real-time engagement metric monitoring, and targeted interventions. IT and telecoms are two examples of technology-driven businesses that encounter particular difficulties with regard to distant work arrangements, virtual collaboration, and digital communication technologies [14].

8.2.1 Encourage Learning

Providing a workforce with opportunities for learning and growth is another way that machine learning can be used to increase employee engagement. Machine learning can be used to pinpoint skill gaps, suggest pertinent courses and design individualised learning programmes for team members. Additionally, machine learning may be utilised to develop interesting and dynamic learning materials that help improve students' knowledge and abilities, such as games, simulations and quizzes. Team leaders may encourage learning so that the team develops, remains current and feels appreciated.

8.2.2 Promotion of Cooperation

Employing machine learning (ML) algorithms to match team members according to their personalities, interests, and skill sets helps organisations build more cohesive and productive teams. This guarantees that projects are staffed with people who have the required experience and fosters a positive and cooperative work atmosphere. Furthermore, regardless of physical location, communication, brainstorming, and problem-solving can be facilitated by employing machine learning to construct virtual collaborative spaces and tools. In today's increasingly dispersed or remote work contexts, where traditional face-to-face interactions may be limited, this is especially crucial.

8.2.3 Boost Well-Being

Enhancing the happiness and well-being of team members is a fourth method by which machine learning can be utilised to increase employee engagement. To help individuals overcome obstacles and enhance their mental health, machine learning can be used to track and evaluate their stress, mood and emotional states. Additionally, companies may utilise machine learning to develop enjoyable and unwinding activities that foster team bonding and relaxation, such as trivia, yoga or meditation. Improving well-being has the potential to increase team morale, motivation and loyalty.

8.2.4 *Exploring Novel Concepts and Being Creative*

To use machine learning to promote employee engagement, first, natural language processing (NLP) is used to analyse survey data and identify common themes and areas for development. Predictive analytics can be used to identify workers who could become disengaged. Experiences can be made more unique by customising training and development recommendations based on ML-driven analysis of user needs and preferences. Gamification components are added to increase motivation by customising challenges and rewards with machine learning. Finally, foster creativity by utilising machine learning to pair staff ideas with relevant partners and resources. Try out these strategies and keep making adjustments in response to comments and outcomes to create a more cohesive and active team atmosphere.

8.2.5 *Suggest Feedback*

Asking staff members for input and acting on it is the fifth technique to use machine learning to increase employee engagement. Machine learning can be used to develop and disseminate questionnaires, polls and surveys that can help team members understand their thoughts, ideas and worries. Additionally, one may utilise machine learning to examine the input and find the major themes, patterns and takeaways that will help to enhance the policies, procedures and leadership. One may demonstrate to the staff that the company values their opinions, experiences and happiness by asking for feedback.

8.3 GenAI

The term "generation artificial intelligence" (GenAI) describes AI systems that can learn and develop on their own using machine learning and deep learning techniques. GenAI involves the use of machine learning algorithms to produce text, images and even code. Its capacity to produce information that appears human has drawn much attention, making it a useful tool in a variety of fields, such as marketing, content development and customer support [13]. The way that businesses handle employee engagement is changing as a result of the use of generative AI. It is critical for HR directors to abreast of emerging market trends and to ensure that the advantages of generative AI are in line with corporate objectives. This most recent AI wave can adapt to various environments and circumstances and requires a wide range of inputs. This demonstrates its value in developing stronger employee engagement methods [15].

GenAI, for example, can automate monotonous chores, giving workers more time to concentrate on important, higher-level work. This increases productivity while

8.3 GenAI

also giving workers a feeling of fulfilment and purpose. Naturally, these factors play a major role in retention and engagement [16]. Additionally, GenAI can predict attrition rates, identify employee pain areas, and direct management to take proactive measures to address these problems through predictive analytics. This has a favourable effect on staff retention and engagement. Furthermore, GenAI is essential for supporting a psychologically healthy workplace culture. To put it simply, GenAI can offer objective, data-driven insights into employee mood and satisfaction, which can encourage more candid, nonstigmatised conversations about difficult issues at work [14].

Employing employee engagement tactics driven by generative AI has the potential to completely transform how businesses interact with their workers. There are several further explanations and possible actions for integrating generative AI into employee engagement:

(a) Beginning with a step-by-step strategy: Gradually integrate generative AI into many HR roles.
(b) Find the main areas that need improvement: Examine current employee feedback and surveys to determine which areas can benefit greatly from generative AI.
(c) The appropriate AI platform is selected based on its features, usability, and compatibility with current HR systems, and different generative AI platforms are compared and contrasted. Scalability, security and compliance are a few things to consider to guarantee successful deployment.
(d) Offer training: Make sure staff members are properly taught to use the new technologies.
(e) Customise the work experience for employees: Make recommendations and personalised content based on the interests and career goals of each employee by utilising generative AI. Depending on each person's needs, appropriate training materials, ideas for skill development and mentorship opportunities can be provided.
(f) Evaluate and assess impact: Keep close eye on and assess how generative AI affects worker engagement. The input from staff members is obtained to determine how satisfied they were with the AI-powered projects. The employee engagement approach was adapted in light of the findings of the evaluation.
(g) Decreased costs: By automating processes and boosting operational effectiveness, generative AI can assist businesses in cutting costs. Over time, this can result in considerable savings.
(h) Enhanced innovation: By offering new concepts and solutions, generative AI may support businesses in becoming more innovative. Increased market share and the creation of new goods and services may result from this.
(i) Data-driven insights for informed decision-making: Generative AI integration transforms decision-making processes by providing organisations with data-driven insights. By utilising AI-generated data, organisations may make strategic decisions that are well informed and in line with their objectives, leading to long-term, sustainable growth.

(j) Global payroll compliance: One of the main benefits of generative AI is its ability to streamline global payroll procedures. This approach decreases the possibility of errors and improves overall payroll accuracy while also guaranteeing cross-border compliance and streamlining intricate payroll processes. A more effective and internationally compliant payroll system is the end result.

(k) Increased employee satisfaction: Using generative AI to streamline activities results in higher levels of engagement and satisfaction among employees. Employees can concentrate on important and difficult facets of their jobs by eliminating pointless complexity, which results in a more satisfying work environment.

(l) Enhanced risk management: By spotting possible issues and offering solutions, generative AI may assist businesses in managing risk. By doing so, businesses may be shielded from monetary loss and harm to their reputation.

References

1. Jha, N., Sareen, P., & Potnuru, R. K. G. (2019). Employee engagement for millennials: Considering technology as an enabler. *Development and Learning in Organisations: An International Journal, 33*(1), 9–11.
2. Garg, R., Kiwelekar, A. W., Netak, L. D., & Ghodake, A. (2021). i-Pulse: A NLP based novel approach for employee engagement in logistics organisation. *International Journal of Information Management Data Insights, 1*(1), 100011.
3. Golestani, A., Masli, M., Shami, N. S., Jones, J., Menon, A., & Mondal, J. (2018, December). Real-time prediction of employee engagement using social media and text mining. In *2018 17th IEEE International Conference on Machine Learning and Applications (ICMLA)* (pp. 1383–1387). IEEE.
4. Nimmagadda, S., Surapaneni, R. K., & Potluri, R. M. (2024). Artificial intelligence in HR: Employee engagement using chatbots. In *Artificial intelligence enabled management: An emerging economy perspective* (p. 147).
5. Medhat, W., Hassan, A., & Korashy, H. (2014). Sentiment analysis algorithms and applications: A survey. *Ain Shams Engineering Journal, 5*(4), 1093–1113.
6. Taboada, M. (2016). Sentiment analysis: An overview from linguistics. *Annual Review of Linguistics, 2*, 325–347.
7. Pasat, A., Birdici, A., & Pop, I. (2021). An internship campaign case study showing results of enhanced recruitment processes using NLP. In *The International Scientific Conference eLearning and Software for Education* (Vol. 2, pp. 222–231). "Carol I" National Defence University.
8. Strang, K. D., & Sun, Z. (2022). ERP staff versus AI recruitment with employment real-time big data. *Discover Artificial Intelligence, 2*(1), 21.
9. Laumer, S., & Morana, S. (2022). HR natural language processing—Conceptual overview and state of the art on conversational agents in human resources management. In *Handbook of research on artificial intelligence in human resource management* (pp. 226–242).
10. Hall, A. N. (2020). *Predicting employee engagement: Machine learning applications to the personality-engagement link* [Doctoral dissertation], Northwestern University.
11. Koo, B., Curtis, C., & Ryan, B. (2021). Examining the impact of artificial intelligence on hotel employees through job insecurity perspectives. *International Journal of Hospitality Management, 95*, Article 102763.

12. Dutta, D., Mishra, S. K., & Tyagi, D. (2023). Augmented employee voice and employee engagement using artificial intelligence-enabled chatbots: A field study. *The International Journal of Human Resource Management, 34*(12), 2451–2480.
13. Rane, N. (2023). Role and challenges of ChatGPT and similar generative artificial intelligence in human resource management. Available at SSRN 4603230.
14. Budhwar, P., Chowdhury, S., Wood, G., Aguinis, H., Bamber, G. J., Beltran, J. R., … Varma, A. (2023). Human resource management in the age of generative artificial intelligence: Perspectives and research directions on ChatGPT. *Human Resource Management Journal, 33*(3), 606–659.
15. Brynjolfsson, E., Li, D., & Raymond, L. R. (2023). *Generative AI at work* (No. w31161). National Bureau of Economic Research.
16. Ooi, K. B., Tan, G. W. H., Al-Emran, M., Al-Sharafi, M. A., Capatina, A., Chakraborty, A., … Wong, L. W. (2023). The potential of generative artificial intelligence across disciplines: Perspectives and future directions. *Journal of Computer Information Systems*, 1–32.

Chapter 9
Predictive Analytics: Anticipating Future Engagement Trends with AI

AI is capable of predicting trends in engagement by using past data and machine learning techniques. It enables active measures for increasing employee engagement and retention, identifying the factors that contribute to an increase in workers' participation and forecasting possible improvement areas. One will be able to unlock the secret patterns, find deeper insight and make critical decisions about creating a workplace that is nurtured by using these AI-powered analytics at disposal.

AI algorithms analyse historical employee engagement data, identify patterns and predict what is to come. It is like having a crystal ball, and it tells us about the future of employee participation [1]. AI can determine the main factors affecting employee satisfaction, productivity and retention through the analysis of previous engagement data. It is possible to find correlations and trends that go much further than surface-level observations. With this knowledge, AI is able to anticipate potential problems in employee engagement, identify areas of improvement and even estimate the number of employees who may be at risk for leaving. Here's the beauty of it all: armed with these predictive insights, one can take proactive measures to prevent disengagement before it even happens. One can implement targeted initiatives, develop personalised interventions, and create a culture that fosters long-term engagement and growth. Unpredictable analytics use AI to predict the future trends and behaviour of customers. This forecasting capacity allows for a more efficient allocation of resources, effective planning of marketing campaigns and quick adaptation to changing market dynamics. Predictive AI analytics enable employees to stay on the front line, avoiding possible errors and exploiting opportunities for improving employee engagement. It is similar to having a strategic compass that leads to a brighter future for employees. Artificial intelligence, which plays a key role in the transformation of raw data into relevant information for marketing purposes, is at the core of predictive analysis. Compared with human datasets, large datasets can be processed and analysed by artificial intelligence algorithms much faster. This speed enables companies to react quickly to changing customer preferences and market trends, which is critical for marketing. Artificial intelligence predictive analytics

© The Author(s), under exclusive license to Springer Nature Singapore Pte Ltd. 2025
S. Majumder and B. Misra, *Analysing Trends and Patterns in Employee Engagement Through AI*, SpringerBriefs in Computational Intelligence,
https://doi.org/10.1007/978-981-96-4496-4_9

allow hype personalisation in all marketing efforts. By analysing customer behaviour and preferences, AI enables businesses to create custom marketing messages and promotions that are effective for each consumer. Artificial intelligence has a role in prediction. Predictive analytics use AI to predict the future trends and behaviours of customers. This forecasting capacity enables marketing managers to allocate resources more effectively, plan effective campaigns and react quickly to changing market conditions.

9.1 Case Study—Applications of Predictive Analytics in Marketing Management

Predictive analytics has a wide range of applications in marketing, which include the following. In this section, a few case studies in which the applications of predictive analytics are described.

9.1.1 Customer Segmentation

Predictive analytics can divide customers into segments based on their characteristics and behaviour, including demographics, psychographics, geographics and firm graphics [2]. This segmentation will allow advertisers to tailor their communications and offers for each group, increasing the likelihood that they will be converted. According to Market Tailor, customer segmentation is an important component of Predictive Analytics and provides a framework for the analysis of customers' data to make precise predictions about how their behaviour will change over time. "Businesses can learn more about their customers, improve marketing strategies and increase growth and profitability through the use of customer segmentation in advanced statistical analysis."

9.1.2 Churn Prediction

"Although it's deceptively simple to calculate turnover with a simple equation—divide customers lost by the total number of customers in a given period—learning how, when and why churn is occurring is more nuanced," according to this Forbes article on how AI can help with churn. "The whole customer experience needs to be seen in a comprehensive way." Data and AI are one way to address this complexity. AI-driven predictive models can identify customers at risk of churning (leaving) and provide insights into the reasons behind their potential departure. This makes it possible for businesses to take active measures to maintain their valuable customers.

9.1.3 Lead Scoring

In business-to-business (B2B) marketing, predictive analysts can assign leads a score on the basis of their likelihood of being sold. For more targeted outreach, sales teams can prioritise high-scoring leads. It is true that sales teams have always performed their own version of lead scoring, but traditional lead scoring is often regarded as too subjective and too dependent on emotion. For decades, businesses have been struggling to prioritise the follow-up of leads. In many cases, salesmen are left to make their own decisions as to who should be contacted first and in the best interests of them. According to HubSpot, marketing and sales professionals use data, such as demographic information, marital status, industry and role, to determine how likely a potential customer is to buy [3]. Those who rank high on this scale are contacted first, while others are contacted last, or if time does not permit, not contacted at all. "To predict future outcomes, predictive lead scoring is based on the use of probability modelling algorithms that analyse data from past clients and current prospects. In HubSpot's view, predictive analytics can help create an ideal customer profile based on past purchasing behaviour and then identify which current prospects are best suited to that profile. It eliminates the possibility of human error or bias, and it relies on hard data to make its predictions."

9.1.4 Content Personalisation

Predictive analytics can generate recommendations for personalised content such as product recommendations, articles or videos to increase user engagement and conversion through the analysis of user behaviour and preferences.

The following are the benefits of predictive marketing analysis. The adoption of predictive analytics in marketing offers numerous benefits: Improved ROI: By targeting the right audience with personalised messages and optimising marketing spending, companies can achieve a greater return on investment for their marketing campaigns. Improved customer experience: If customers receive content and services that are of great relevance to their needs and preferences, personalised marketing based on predictions will lead to better customer experiences. Competitive advantage: By staying on top of market developments and reacting to changes more quickly than their competitors, companies using predictive analytics are gaining a competitive advantage. Cost savings: Predictive analytics can help companies allocate resources more effectively and reduce unnecessary expenditures by identifying inefficiencies in marketing campaigns. Better decision-making: Data-driven understandings from predictive analytics authorise vendors and decision-makers with the information they need to make informed and tactical selections [4].

9.2 Predictive Analytics Tools for Marketers

A wide range of powerful tools and technologies are used for predictive marketing analysis. Valuable information from vast datasets can be obtained from data mining software such as RapidMiner and KNIME. The development of predictions is made possible by machine learning libraries such as TensorFlow and Scikitlearn. Customer relationship management (CRM) systems, including Salesforce and HubSpot, aid in customer data management and segmentation. Predictive intelligence and content are used by marketing automation platforms such as Marketo or HubSpot to carry out campaigns. "With its free and low-priced entry-level plans, Hub Spot works well for small businesses and solo entrepreneurs," according to a Forbes comparison of the two. It is a good choice for medium to large businesses with CRM systems in place, Salesforce or Microsoft Dynamics 365 yet need an effective marketing automation tool that can handle complex advertising campaigns. The deployment and implementation of predictive models is facilitated by cloud-based analytic solutions such as AWS SageMaker and Google Cloud Artificial Intelligence. These tools collectively empower marketing managers to make data-driven decisions and improve the effectiveness of campaigns.

9.3 Caselets on Employee Engagement in Various Organisations

An essential component of any successful organisation is the involvement of employees. Employees who are engaged in their work have a higher level of productivity, innovation and commitment. It is also more likely that they will continue to work for their existing employers, which will help to reduce turnover rates. According to a Gallup survey, only 36% of US workers are employed. However, companies such as Google have proven that 20% of productivity gains can be achieved by increasing engagement. Google has established a standard for effective staff involvement through the implementation of innovative strategies, such as flexible working conditions and continuous feedback. In this section, several cases on employee engagement from various companies' perspectives are presented.

9.3.1 Google: The Incentives for Employee Engagement

Google, a renowned technology giant, has consistently been a leader in promoting employee engagement via creative methods and a nurturing workplace. Google's "20% Time" initiative is one of the most prominent policies that has attracted worldwide attention. This means that staff can spend one-fifth of their working time on personal projects chosen by them. There was a marked increase in both engagement and innovation. Several of Google's best-selling items, such as Gmail and Google News, originated from its 20%-time policy. By allowing employees' independence and self-governance to pursue their passions, Google has discovered a creative resource that helps with product development and business expansion in addition to providing benefits to staff members. This initiative clearly signals to employees that their concepts and goals are valued, encouraging a feeling of responsibility and engagement with their work. In addition, Google is investing significantly to create an environment where employees are well cared for and have a strong sense of identity [5]. Free meals, on-site medical care and relaxation areas are some of the enticing perks offered by the firm. In reducing stress and improving the quality of working life for employees, these benefits go a long way.

This emphasis on well-being is not merely a gesture of charity but also a deliberate decision. Google recognises that satisfied and healthy staff are more engaged and productive. Employees who feel valued are more likely to remain with the company, lowering turnover and maintaining important talent. Google has continuously maintained high levels of employee engagement and retention by sticking to initiatives such as the 20%-time initiative and maintaining an enriching work environment. The company has established a standard for the technology industry, demonstrating how long-term investment in employees' skills and well-being pays off.

9.3.2 Zappos: Enlightening an Energetic Company Culture

An excellent example of how a distinct company culture can contribute to employee engagement is Zappos, an online retailer specialising in footwear and clothing. The firm commitment to a set of core values that define Zappos' culture is at the heart of its remarkable success story. Two of the most pivotal values are "Deliver WOW through Service" and "Create Fun and A Little Weirdness." Zappos's dedication to integrating these values into every aspect of the company, beginning with the recruitment process, truly sets it apart. In assessing potential hires, Zappos is strongly interested in the compatibility of culture with job skills. In essence, this culture fit ensures that employees are genuinely satisfied with the values and mission of the company.

Two outcomes are evident from this cultural integration. First, it creates an environment where workers discover a strong sense of meaning and purpose in their job. Work satisfaction and engagement come easily when they are in line with core principles. At Zappos, employees do not merely process orders; they actively contribute to delivering WOW to customers and thrive in an environment that embraces a touch of weirdness. Second, a common ethos and sense of belonging within the workforce are brought about by this focus on cultural fit. Zappos employees are not just working for the company; they are working on behalf of their cause, surrounded by colleagues who share their passion [6]. Zappos has become the gold standard for employee engagement, showing a clear link between its unique company culture and employee satisfaction. By cultivating a dynamic corporate culture that prioritises values and cultural compatibility, Zappos highlights the importance of recruiting individuals who align with the company's core values. This alignment has proven to enhance engagement, motivation and overall workforce contentment, as evidenced by their success.

9.3.3 Microsoft: A Data-Driven Attitude

Microsoft is a giant in the technology world, and its innovative approach to employee engagement stands out among others. Through its "One Microsoft" programme, the company has embraced a data-driven approach, and the results demonstrate the power of listening to employees. Microsoft's strategy revolves around receiving employee feedback on a regular basis. This feedback is more than simply a formality; it is an essential tool for making strategic adjustments. By constantly seeking employee feedback, Microsoft signals that their thoughts and concerns are valued. One of the most impressive outcomes of this data-driven strategy is the overhaul of Microsoft's performance review process. In the past, performance reviews were dreaded by many companies, often focusing on rankings and evaluations. Microsoft, however, made a shift in approach by taking into account the feedback of its employees. The organisation realised that focusing on employee engagement was more advantageous to the performance evaluation process, emphasising development and growth over competition. Microsoft's revised method encourages collaboration and skill development, removing the stress of employees competing against one another. Instead, they are encouraged to collaborate and assist each other's professional development. This move has resulted in a more positive work atmosphere, inspiring individuals to succeed by ensuring that their efforts are recognised and that they have possibilities for personal and professional development. Microsoft's data-driven strategy exemplifies a crucial element of employee engagement: actively listening to and responding to feedback from employees. By demonstrating how employee perspectives can have a significant impact, Microsoft has fostered an atmosphere of trust and collaboration, increased employee involvement and benefiting both the firm and its workforce.

9.3.4 Hilton: Gratitude and Obligation

Hilton is a shining example of how recognition and gratitude can help keep employees engaged and increase corporate success in the highly competitive global hotel industry [7]. Hilton recognises the critical role that motivated and dedicated workers play in giving visitors a memorable experience. To promote this involvement, Hilton has implemented a number of programmes and processes that acknowledge and reward outstanding employee performance. 'Catch Me at My Best and the Spirit of CARE' are two key campaigns. The Catch Me at My Best programme allows employees to recognise and appreciate their coworkers' remarkable efforts and accomplishments. This approach promotes a peer-to-peer recognition culture in which staff members support one another in addition to working for the organisation. Recognition leads to a feeling of appreciation and creates an atmosphere in which workers are respected for their commitment and hard work. Another sign of Hilton's commitment to employee engagement is the "Spirit of CARE" programme. This programme will reinforce the idea that employees tend to be more attentive towards their guests when they look after each other. The importance of teamwork and shared values is stressed in fostering an uplifting, cooperative workplace within the Hilton community.

The results of a focus on recognition and appreciation at Hilton were outstanding. The direct result of having engaged staff who are proud of their work is improved customer satisfaction. It is more likely that workers will go far beyond to guarantee a remarkable experience for their guests when they are treated with respect and appreciation. In addition, the increase in employee retention rates at Hilton points to a relationship between staff engagement and business success [8]. It is more likely that an engaged staff member will remain in the company, reduce turnover costs and ensure a level of service consistent with quality standards. A notable case study in the hotel industry is Hilton's emphasis on recognition and appreciation through programmes such as "Catch Me at My Best" and "Spirit of CARE." These initiatives highlight the importance of cultivating a culture where employees feel valued and motivated, significantly enhancing customer satisfaction and driving business success. Hilton has proven to be an effective brand ambassador, driving the success of the company as a whole when employees are involved and encouraged.

9.3.5 The Ritz Carlton: Endowing Employees to Drive the Additional Mile

Within the hospitality sector, Ritz Carlton is a prime illustration of how putting staff engagement first can result in a superior client experience. By introducing an environment of autonomy and trust among employees and providing staff with the power to make decisions that benefit guests who do not require management approval, the reputation of a hotel chain has strengthened. The level of autonomy given to staff is one of the cornerstones of that strategy. A unique culture of trust has been created by Ritz Carlton, where employees are not only seen as workers but also considered to be valued decision-makers. This freedom allows staff to make use of opportunities for improving the guest experience, whether it is a free upgrade, personalised equipment or resolving an issue on time. It is not only a culture of trust that this level of autonomy creates but also an intense sense of employee involvement [9]. Employees feel a greater sense of ownership over their roles when they are entrusted with this responsibility, leading to more committed and motivated workforces.

Exceptional customer experiences provided by The Ritz Carlton clearly illustrate the results of this approach. There is a consistently high level of employee satisfaction in the case studies for various Ritz Carlton properties. Excellent customer service is the consequence of contented staff members who are motivated to go above and beyond to ensure that visitors have an unforgettable experience. The positive word-of-mouth that results from these exceptional experiences is priceless in a field where reputation is everything. In addition, the bottom line is positively impacted by this employee-centred strategy. An increasing number of guests are returning, giving the hotel a favourable review or recommending it to others. This is reflected in increased revenues and profitability for Ritz Carlton, reinforcing the positive correlation between employee involvement and business success. An excellent example of how priorities for engagement can enhance customer service to an exceptional level is Ritz Carlton's approach toward empowering employees. Through the establishment of a trusting culture and the delegation of authority, the organisation has developed a devoted and motivated staff that is committed to providing outstanding guest experiences. The achievements of Ritz Carlton provide a fascinating case study that shows how, in the hospitality industry, employee engagement can be the cornerstone of corporate greatness.

References

1. Koo, B., Curtis, C., & Ryan, B. (2021). Examining the impact of artificial intelligence on hotel employees through job insecurity perspectives. *International Journal of Hospitality Management, 95*, Article 102763.
2. Othman, S., & Mahmood, N. (2019). Linking employee engagement towards individual work performance through human resource management practice: From high potential employee's perspectives. *Management Science Letters, 9*(7), 1083–1092.
3. Lemon, L. L. (2019). The employee experience: How employees make meaning of employee engagement. *Journal of Public Relations Research, 31*(5–6), 176–199.
4. Clack, L. (2021). Employee engagement: Keys to organisational success. In *The Palgrave handbook of workplace well-being* (pp. 1001–1028).
5. Bhattacharya, A., & Majumder, S. (2023). Predictive analysis on absenteeism at a workplace using explainable AI. In *PaKSoM 2023* (p. 63).
6. Ghosh, S., Majumder, S., & Peng, S. L. (2023). An empirical study on adoption of artificial intelligence in human resource management. In *Artificial intelligence techniques in human resource management* (pp. 29–85). Apple Academic Press.
7. Majumder, S. (2022). AI in health and safety management for real estate 4.0. *International Journal of Ambient Computing and Intelligence (IJACI), 13*(1), 1–18.
8. Kelley, S. (2022). Employee perceptions of the effective adoption of AI principles. *Journal of Business Ethics, 178*(4), 871–893.
9. Mukherjee, S., & Agrawal, P. (2023). Disengagement to engagement: A strategic employee retention approach with AI. In *The role of HR in the transforming workplace* (pp. 147–155). Productivity Press.

Chapter 10
Ethical Considerations and Future Directions of AI in Employee Engagement

The ethical implications of adopting AI must be taken into account by organisations, given the number of opportunities for employees to engage. A critical consideration is the balancing of privacy and surveillance in AI monitoring systems. In the use of AI and employee analytics, organisations need to ensure transparency and credibility to mitigate bias and guarantee fair treatment. In addition, the creation of AI systems that foster employee trust and collaboration is essential for maintaining a sound relationship between people and computers. Navigating ethics issues and considering the future direction of artificial intelligence at work is essential, as organisations adopt AI to engage their staff [1].

10.1 Ethical Use of AI for Employee Monitoring

AI-powered employee monitoring systems present serious privacy and security concerns. To maximise efficiency, organisations must strike a balance between monitoring employee behaviour and protecting personal privacy. Creating ethical AI-powered employee analytics solutions that foster a healthy work environment necessitates openness and trust. AI monitoring systems are rapidly gaining popularity, with astonishing features such as detecting keystrokes and browsing trends, as well as forecasting employee turnover. However, these tools also raise serious privacy concerns. Beyond legal concerns, there is a substantial psychological impact on employees. A constant surveillance culture can heighten worry and stress, diminishing the productivity that these instruments are intended to boost [2]. It can also impede innovation since employees who feel continually monitored are less likely to take risks or think innovatively.

© The Author(s), under exclusive license to Springer Nature Singapore Pte Ltd. 2025
S. Majumder and B. Misra, *Analysing Trends and Patterns in Employee Engagement Through AI*, SpringerBriefs in Computational Intelligence,
https://doi.org/10.1007/978-981-96-4496-4_10

10.2 Ensuring Human-AI Collaboration and Empathy

However, human interaction continues to be crucial in the context of AI's vast possibilities for employee engagement. In view of the importance of emotional intelligence, organisations need to develop systems for artificial intelligence that foster trust and cooperation between employees. Fostering empathy for AI assistance ensures supportive and humanlike interactions that increase employee engagement and satisfaction. In human AI collaboration, empathy involves AI systems that understand and respond to human emotions. Artificial intelligence may analyse a user's voice tone or facial expressions to determine his or her emotional state, e.g., in the case of virtual assistants. The aim is to create a more natural and intuitive interaction between humans and machines through this empathy approach. For example, a system of AI called HAILEY has been developed to provide humans with immediate feedback on their behaviour and help them respond more empathically [3]. The collaboration between humans and AI in various areas has manifested. In the healthcare sector, AI algorithms help doctors diagnose diseases by analysing large sets of data, such as medical images and patient records. This not only helps to speed up the diagnosis process but also improves its accuracy. To confirm the findings, interpret complicated cases and make definitive diagnoses, radiologists can use artificial intelligence algorithms. In a number of sectors, including customer services, healthcare and psychiatric care, AI has been applied to collaborate with people.

Human-AI collaboration is more common in low-risk tasks such as creating emails with Gmail's Smart Compose for real-time writing assistance or checking spelling and grammar. Augmenting human efforts with AI, rather than replacing them, is critical, especially in areas that require human judgement and oversight, such as complicated mental health care jobs and high-risk military decisions.

10.3 Exploring Emerging Trends and Innovations in AI for Employee Engagement

There are exciting possibilities for the future of artificial intelligence with regard to employee engagement. Enhanced interactive environments that encourage employees' engagement and collaboration can be created by the integration of AI with augmented reality. In addition, neuroadaptive artificial intelligence provides an opportunity for personalised and tailored employee experience. Organisations can unlock the real potential of AI technology if they develop strategies for future employee engagement based on AI technologies. In the field of predictive analytics, artificial intelligence also plays an important role in enabling organisations to anticipate and respond to engagement issues in a timely manner. Artificial intelligence algorithms can identify patterns and trends through the analysis of vast amounts of data, which can help organisations identify potential withdrawal factors before they become a major problem. For organisations that seek to attract, retain and nurture top talent in a competitive environment, the next generation of employee experience is crucial. The creation of a single, consumer-grade experience for the organisation's workforce through the use of digital technologies will have a positive impact on all human resource processes and the dimensions of the employee's work life [4]. The core of the HR officer's agenda and a top priority for organisations today continues to be employee experience EX.

Compared with those who do not, companies investing in EX have a compelling value proposition. They have been shown to generate four times higher average profits and twice as much revenue. Moreover, they are 11 times more likely to be featured on employee review sites as best places to work and more than two times as often among the world's most innovative companies. In addition, according to EY's analysis, their teams are 21% more productive, and employees are 60% more likely to be retained by their employers.

The HR technology market is lively, and thousands of suppliers are now active as the market continues to evolve. More than $4 billion in venture capital and approximately 300 funding rounds have been reported for the sector since 2023. Moreover, according to various reports, more than 200 mergers and acquisitions have taken place in this sector. All aspects of human resources and enterprises are being revolutionised by artificial intelligence. CoPilots are abundant, and every aspect of AI is evolving exponentially, with enormous business impact, as GenAI enters the market. While individual AI and digital dimensions are powerful on their own, they are even stronger with each other. One example of how technology works well in synergy and builds upon each other is conversational AI (NLP). The combination of GenAI, ML and analytics is another example. There are a lot of combinations out there.

10.4 AI Integration with Current Systems

It can be difficult to integrate AI with current procedures and systems. AI techniques for seamlessly integrating systems include using APIs and middleware, finding integration points, and carrying out in-depth system inspections. Organisations can optimise the efficiency and efficacy of artificial intelligence implementation by guaranteeing compatibility and scalability.

10.5 Making AI Ethical and Objective

While using AI, bias and fairness are crucial factors to consider. One should look at methods for spotting and removing biases in AI systems and make sure that judgement calls are made fairly. To address ethical problems, organisations should establish frequent audits and reviews and encourage inclusiveness and diversity in AI development projects.

10.6 Constant Monitoring and Assessment

To pinpoint problem areas and guarantee continued efficacy, monitoring and assessment of AI systems are crucial. One should discuss how crucial it is to set up performance indicators, carry out frequent evaluations and obtain input from stakeholders. Organisations may optimise and develop AI solutions by regularly monitoring and assessing their performance [5].

10.7 Handling Information Security and Morality

Because artificial intelligence (AI) uses large amounts of data, privacy, security and ethical issues are important to address. People will examine the significance of data governance, guarantee adherence to pertinent rules and put robust safety measures in place. People will also talk about the ethical ramifications of AI and the necessity of responsibility, justice and transparency in AI systems.

10.8 Handling of Employment Issues

The introduction of AI technology frequently causes people to worry about their jobs and how their functions will be affected. People will investigate methods for resolving these issues by emphasising AI's function as an augmentation rather than a replacement, offering opportunities for improvement and training, and maintaining open lines of communication. Through employee involvement in the implementation process and resolving their worries, organisations can cultivate a climate conducive to the widespread implementation of AI.

10.9 Expansion of Execution

AI solution development and implementation can be costly, particularly for smaller companies. Affordable methods are required to make AI available.

10.10 Responsibility and Disclosure

AI may operate as a "black box," making it challenging for employees and HR specialists to comprehend the decision-making process. Companies that provide a clear explanation of how AI algorithms make judgments might boost employee confidence in the technology.

10.11 Low Emotional Intelligence

AI finds it difficult to comprehend the human element of human resources. It is unable to replace emotional intelligence, empathy, or the capacity to manage difficult negotiations or disagreements at work. HR specialists continue to be essential in fields where these abilities are necessary.

10.12 Antiquated Facilities

AI systems must process vast volumes of data in a matter of milliseconds to achieve the desired outcomes. Operating on devices with appropriate infrastructure and processing power is the only way to accomplish this goal. Nonetheless, many companies continue to operate with antiquated technology that is completely incapable of handling the demands of implementing AI. It follows that companies seeking to use machine learning to transform their learning and development processes need to be ready to make investments in cutting-edge hardware, software, and infrastructure.

10.13 Complexity of the Learning Curve

A learning curve is introduced by the integration of AI as HR personnel adjust to new technology. Programmes and materials for training employees are crucial to ensuring that they know how to apply AI tools in their regular work.

10.14 Organisational Culture Shift

The adoption of AI requires a change in organisational culture. This entails fostering an environment where workers view technology as a tool to improve their work and encourage cooperation between AI-driven and human operations.

10.15 Combining Services for Generative AI

There are particular difficulties with generative AI services since they require the creation of new data or content. Companies need to carefully evaluate aspects such as originality, data quality and potential ethical ramifications when integrating generative AI into HR procedures.

Apart from the difficulties in implementing AI, differences in global AI availability can also be explored [6]. In particular, some nations are already advancing significantly in AI technology, while others are finding it difficult to catch up with much more basic technological developments. Furthermore, because artificial intelligence occasionally requires data that are subject to data protection rules, there are a number of ethical and legal issues surrounding this technology. Numerous discussions have already taken place to establish legislation that will guarantee security and transparency [7].

Businesses, governments and other organisations must overcome the many obstacles that AI deployment presents if they are to reap its benefits and participate in the development of machine learning in the future [8]. Hopefully, the ambiguity concerning AI will gradually fade as more research is conducted on it.

References

1. Budhwar, P., Malik, A., De Silva, M. T., & Thevisuthan, P. (2022). Artificial intelligence—Challenges and opportunities for international HRM: A review and research agenda. *The International Journal of Human Resource Management, 33*(6), 1065–1097.
2. Majumder, S., & Dey, N. (2023). Artificial intelligence: The future of people management. In *The vogue of managing people in workplace* (pp. 83–102). Springer Nature Singapore.
3. Sachan, V. S., Katiyar, A., Somashekher, C., Chauhan, A. S., & Bhima, C. K. (2024). The role of artificial intelligence in HRM: Opportunities, challenges, and ethical considerations. *Educational Administration: Theory and Practice, 30*(4), 7427–7435.
4. Tippins, N. T., Oswald, F. L., & McPhail, S. M. (2021). Scientific, legal, and ethical concerns about AI-based personnel selection tools: A call to action. *Personnel Assessment and Decisions, 7*(2), 1.
5. Majumder, S., & Dey, N. (2022). Explainable artificial intelligence (XAI) for knowledge management (KM). In *AI-empowered knowledge management* (pp. 101–104). Springer Singapore.
6. Okatta, C. G., Ajayi, F. A., & Olawale, O. (2024). Navigating the future: Integrating AI and machine learning in HR practices for a digital workforce. *Computer Science & IT Research Journal, 5*(4), 1008–1030.
7. Zhang, Y., Xu, S., Zhang, L., & Yang, M. (2021). Big data and human resource management research: An integrative review and new directions for future research. *Journal of Business Research, 133*, 34–50.
8. Ooi, K. B., Tan, G. W. H., Al-Emran, M., Al-Sharafi, M. A., Capatina, A., Chakraborty, A., … Wong, L. W. (2023). The potential of generative artificial intelligence across disciplines: Perspectives and future directions. *Journal of Computer Information Systems*, 1–32.

Conclusions

A number of industries have been revolutionised by AI, and the area of employee engagement is no exception. Today, organisations recognise the potential of artificial intelligence to improve employee morale, productivity and satisfaction at work in a rapidly evolving professional environment. Organisations can use AI's capacity to increase the engagement and motivation of their workforce by understanding the role it plays in employee engagement, examining its benefits and taking ethical considerations into account when making adoption decisions. Various aspects of using AI to improve recognition, reward, communication and collaboration as well as the development and learning of employees have been discussed in this extensive guide on how to use AI for staff engagement. Organisations can promote a more engaged workforce, unlock growth potential and enhance overall employee satisfaction by integrating intelligent AI and ethical practices. It is not only a trend to embrace AI's potential for employee engagement but also a way of strategically building a workplace that harnesses innovativeness and enables employees to be successful.

The manufacturer's authorised representative in the EU is Springer Nature Customer Service Centre GmbH, Europaplatz 3, 69115 Heidelberg, Germany. If you have any concerns regarding our products, please contact ProductSafety@springernature.com

Printed and bound by CPI Group (UK) Ltd, Croydon, CR0 4YY

26/03/2026

02078983-0002